DEOVOLENTE

A.L. Butler II

ISBN: 978-1-961445-84-0

Table of Contents

Dedication

To my dad, even when doubt crept into my mind, you were there, reminding me of my talent and potential. You taught me the value of storytelling and instilled in me the courage to share my voice with the world. Thank you, Dad, for inspiring me, encouraging me, and always reminding me that my words have the power to touch hearts and minds. Your love and guidance have made all the difference. This book is dedicated to you, as a token of my deepest gratitude and love.

To my loving wife, whose belief in my dreams never wavered. You have been my anchor, my muse, and my constant source of encouragement. Your unwavering support and understanding made it possible for me to bring this book to life. Thank you for standing by my side throughout this incredible journey.

And to my precious children, whom I was absent from during their childhood. My unconditional love for all of you has been my driving force. This book is dedicated to you, as a reminder that dreams are worth pursuing and that the power of imagination knows no limits.

May this book be a testament to the profound impact you have all had on my life. Your love, guidance, and unwavering belief in me have made all the difference.

Preface

A thrill-seeking young man with ambitions and hope for a bright future has his world upside down when he lands in prison. His dedication won't let him give in. He continues to pursue his passion from behind the bars. Akbar was unstoppable, and nothing could stop him from pursuing his education. He utilized his time in jail, which others would perceive as a curse, to teach himself computer programming.

His passion led him to continue his studies in robotics. He broke the barriers of his dark past, which was weighing him down. But can someone really leave their past behind? The answer for Akbar was clear, as his past was chasing him.

He had yet to disclose a secret from a heist he had committed with his best friend years earlier. This secret had made its way back to haunt Akbar. His worst nightmare became a reality when his father and best friend were murdered.

Akbar's passion had reshaped into something that kept a flame burning in his heart. He was out to seek revenge for the murder of his father and best friend. He had a lot of questions, but he was yet to find the answers.

With a vengeance in his heart, his hands were ready to get stained with the blood of his father's murderer. Akbar was out to find the killers and ensure justice is served. But he had to find answers first:

- Who killed my father and my best friend and why?
- What was the secret that led to them being killed?
- Where could he find the clues to trace his father's killer?

This is the story of Akbar's revenge. They say, "*What goes around always comes around.*" Akbar had embarked on the journey of avenging the brutal killings of his loved ones. Helping him were two of his self-built humanoids.

Chapter 1: Deo Phase 1

"Damn!" Janice yelled out.

Quick looked over at Janice, directly into her eyes. They both knew what had just happened; they played Russian roulette while having unprotected sex. Quick and Janice had both received full scholarships to top universities. Janice was on an academic scholarship for pre-med at UCLA and Quick at Texas A&M on a baseball scholarship.

"Janice, I told you we should have used protection," Quick said. "We cannot afford to get pregnant; we are both being foolish!"

Janice replied, "Baby, forgive me for pressuring you into having unprotected sex."

Quick, realizing his mistake, he tried to make up for it. He caressed her face lightly, holding it by the chin as he passionately kissed her.

"Quick, I love you; I would never abandon you," Janice replied.

Quick reached over for his pants, eyes still locked with Janice's. They were the same. Their fathers were childhood friends, too, both having served prison terms together. Janice's father was still serving a life sentence, and Quick's father had served time for manslaughter, which his conviction was later overturned on an appeal. So, they both understood the pain of not having a parent by your side.

Janice took martial arts classes at an incredibly early age and became one of the top ten junior karate champions in her region. For Janice, martial arts were her therapy. It became a way for her to vent her pain. Quick and Janice were, like, a dynamic duo. Both were highly intelligent and highly motivated to become successful and productive citizens. But they also knew what it felt like to be alone and learned to treasure one another.

All Quick and Janice had was each other. They would smile when they spoke, knowing what the other had to say before they began. This happened a lot with these two. The warmth radiated from their love warmed the entire vehicle when they were driving out.

Janice said, "Quick, we need to hurry up. My Aunt Kay will be working early."

KNOCK KNOCK.

Janice banged on an all-black iron door. You could hear Ed Sheeran's song "Thinking out loud" as it blared from within the house. Janice knocked once more, and before she could another time, she heard the voice Aunt Kay yelled out.

"I am coming," she attempted to yell over the music playing.

Suddenly the massive iron door swung open, almost hitting Janice in her head.

Aunt Kay said, "I am so sorry, baby. Come on in. I am running late." Aunt Kay ordered J.C. to turn the music down.

J.C. did accordingly, and then he noticed his cousin Janice standing next to his mother. He smiled at Janice.

J.C. said, "Hello Janice, I have missed you; how have you been?"

Janice replied, "I have been doing well."

J.C. said, "A friend of mine broke his arm playing baseball today."

"Wow, J.C., every time I see you, something happens to one of your friends," Janice remarked.

J.C. walked toward Janice and hugged her.

Janice replied, "Damn, damn, J.C.! Don't squeeze me to death!"

They both laughed. J.C.'s voice was too deep for his age. He sounded like a much-older man—always.

J.C. said, "You already know you are my favorite cousin, Janice. I love you."

Aunt Kay said, "Well, I am on my way out!"

J.C. and Janice made their way toward the kitchen. Janice couldn't help but worry about Quick with him being out on the streets. It was well known that his family had endured constant harassment by the Long Beach Police Department in the neighborhood. At that moment, her phone started to ring.

Before she could reach the refrigerator, Janice reached into her pocket and noticed it was her mother calling.

Janice answered, "Hello, mom!"

"Hello Angel, will you be home tonight?" said her mother.

"Most likely not, Mom, Aunt Kay will be working overtime, so I'll stay the night. I have plenty of clothes here."

Janice's mother replied, "I am so glad I called. I do not have to worry or cook anything before my shift."

Janice said, "Okay! Mom, I'll talk to you tomorrow."

J.C. yelled, "Come on, Janice, let's play Call of Duty!"

Janice hung up and made her way to J.C.'s bedroom to play Call of Duty as they both played throughout the night. They both had managed to fall asleep, game controllers in their laps.

***

Boom, a loud explosion was heard, and the sound vibrated through the entire house.

Janice had been awake and looking at her watch that read 5:30 A.M. She looked at J.C., who was still asleep.

J.C.'s bedroom blinds had been open, and a light breeze was coming through. The excruciating sound made Janice's ears ring. She was surprised that J.C. had not woken up. Janice rubbed her eyes; she could see an orange light illuminating J.C.'s bedroom window. She could smell smoke coming through.

Janice yelled, "Get up. Get up, J.C.."

Janice did not know if something in the home had caught on fire, but she was not taking any chances. J.C. finally woke up, and they began to move their way out.

"Stay behind me," Janice instructed J.C. as they exited the bedroom.

Janice's ears were still ringing. She scanned the house, making sure everything was in order. Suddenly Janice heard a voice and the sound of someone banging on the door.

"Hurry! Hurry! Please come out; it is Kay!" A deep voice yelled out from behind the iron door.

Not knowing what had just happened, Janice had given instructions for J.C. to grab her cell phone.

But before J.C. could retrieve it, the voice behind the door shouted, "It is me, Smitty, from across the street! It is Kay—her car exploded!"

J.C. gave Janice her phone, but she didn't use it. She couldn't. After hearing what had come out of Smitty's mouth, Janice opened the door.

Aunt Kay's car was on fire.

She looked back to see where her cousin was, as she had instructed J.C. to stay behind her. Knowing that if J.C. saw the lit vehicle, he would be traumatized.

Smitty stood there with tears in his eyes. Janice stood on the front porch; she could smell burnt rubber and something else so horrible that she began to vomit. She looked toward her neighbor's yard and noticed an arm that the fire had charred.

You could see what had happened.

It appeared to be a hand. Janice could not believe she had been so abruptly awakened to chaos. Janice focused her eyes. She heard a scream.

As Janice turned, J.C. was standing there with tears and in shock. Janice walked toward J.C. to console her cousin. You could hear sirens in the distance. J.C. continued to look over Janice's shoulder as Janice hugged J.C. tightly.

Janice said, "No, baby, do not look back!"

Janice knew the damage this loss would cause. Having just woken up to such an explosion and seeing your loved one being burned alive was a big deal. It was a mess.

"Let me go, you fucking bitch," J.C. yelled angrily.

Janice understood that J.C. did not mean what he said.

Janice suddenly got a phone ring in her hand and answered it, "hello."

On the other end of the phone, a woman's voice said to Janice, "How did you like those fireworks?"

This person was laughing—maniacally, pure evil. Then the call went silent, and finally, it disconnected. Janice stood there in shock. She then heard a voice coming from behind her.

"Excuse me, miss," the voice spoke out.

As Janice looked back, standing behind her with a pen. And notepad in hand stood a Long Beach police officer.

"Sorry to scare you?" The officer said.

"What do you want?" Janice growled.

"Excuse me, but can I ask you a few questions?" He pointed toward Aunt Kay's home.

Janice did not want to look at the officer, knowing she could see her aunt's burned, charred remains.

The three went inside as the officer touched Janice softly on her right shoulder. Neighbors stood outside their homes, crying and shaking with disbelief, not believing what had just happened on their block.

The officer then escorted Janice and J.C. toward the front door of Aunt Kay's home. Janice and J.C. held each other's hands as they walked into the house. As they entered, the officer directed Janice and J.C. toward the couch.

As they sat down, the officer introduced himself, "Hi, I am officer Randall; I am sorry, but I heard you are the victim's family."

Suddenly, they heard the voice of footsteps approaching.

Officer Randall turned to see who it was; his partner, Officer Conway.

Officer Randall introduced him, "This is my partner, Officer Conway."

Janice looked very irritated with them, thinking, what could I possibly tell them?

Officer Randall, "Did you see or hear anything?" He asked.

Janice replied, "All I know is that I was awakened by a loud boom in which the whole house vibrated."

Officer Conway looked around the house as Randall continued with his questions.

Officer Randall asked, "So, who are you to the victim?"

Janice replied, "She is my aunt, and this is her son, Jason."

Officer Randall said, "Okay, that is all for now!"

Officer Randall put his pen and pad, knelt in front of Janice and J.C., and said, “I am sorry for your loss, and I will keep you in my prayers.”

Officer Conway returned with a bottle of water and Gatorade and handed one to each.

Officer Conway asked Janice, “So, what is your name?”

Janice replied, “I am Janice Kemp…” she faltered. “That’s… that’s my aunt in the car.”

J.C. began to cry at the mention of his mother.

Officer Randall tapped Officer Conway on the shoulder and said, “Let’s go; they have been through a lot this morning.”

Officer Randall looked down at Janice and politely inquired, “Is there anyone you could call before we go?”

Before Janice could respond, a call came in on her phone. It was Janice’s mother;

“Baby. Are you okay?”

Janice replied, “Yes, mom, I am fine. Two officers asked me a few questions, and they are leaving now.”

Janice’s mother said, “I am on my way, baby. I am coming down the 710 freeway right now, and I should be there in 15 minutes.”

Janice began to cry over the phone. Janice looked up at Officer Randall and Conway and said, “My mother is on the way.”

In a gentle voice, Officer Randall reached into his pocket and said, “Here is my card; call me if you need me.”

The card read—TJ Randall 562-122-6115, Southern Division.

Officer Randall knelt beside Janice and J.C., looked at them in their eyes, and told them to be strong. Janice and J.C. were still in shock.

In the distance, you could hear fire trucks and police cars arriving. Neighbors were still having conversations about the tragedy on their block. J.C., no longer crying, stared shocked at the grandfather clock in their dining room. Janice wanted to talk to J.C., but as she looked into his eyes, she could see a look of despair had overcome him.

She took her cousin's hand and said, "Come on, baby, let's go and lay down."

They headed toward J.C.'s room, and emptiness could be felt throughout the house, knowing that Aunt Kay would never return. As they approached J.C.'s room, the phone rang. She looked at the incoming call.

She managed a smile, albeit a weak one. It was Quick.

Janice answered the call, "Hello, baby!"

Quick asked, "Janice, are you okay? I will be there in a few minutes. Just hang tight."

Janice replied, "I am not feeling well, but all I can do is stay calm for J.C.."

She looked over at J.C. and hung up the call.

Janice still could not believe that Aunt Kay had been blown to pieces. The vision of her limbs scattered throughout the street was stuck in her head. Janice still did not know how much J.C. had seen. As Janice and J.C. lay down, she fixed her hair, tying it in a bun.

She embraced her cousin and whispered to him. "It will be okay, J.C.."

J.C. was quiet.

They both lay waiting for Janice's mom to arrive, but their block had been yellow-taped. Janice's mom approached the neighborhood but could not get close to Aunt Kay's home as it was barricaded for about five to six blocks. She drove around looking for a parking spot. Finally, she located one on 20th and Lemon. She noticed ATF jackets on several individuals standing around. She still could not believe what was happening. She grabbed her phone and called Janice.

On the first ring, Janice answered, and it startled her. It seemed as though she had only laid down for a brief moment.

Janice's mother said, "I am a few blocks down, and they have the entire street taped off. Hopefully, I will surpass these few federal agents, and regular police are everywhere!"

"Okay, J.C., and I will come to you."

Janice's mother replied, "I am parked on 20th and Lemon."

As they hung up, Janice said to her cousin, "J.C. put your shoes on; my mom's here." Her voice cracked.

They were still in shock. They couldn't forget what they had witnessed earlier in the morning, with Aunt Kay being killed. Janice and J.C. made it past a few agents as they walked outside.

Suddenly, a tall, thin, red-haired agent called out, "Excuse me, but nobody can leave outside this perimeter," he explained.

Janice said, "Sorry, Sir, but the person in that vehicle that exploded was my auntie and my cousin's mother."

"Well, until this area is clear, no one can exit," the agent countered.

Trying to keep her calm, Janice replied, "Well, Sir, my mother is trying to get to us."

"Sorry for your loss. I am an agent Zorokowski with the ATF. I am the lead agent on this case. I will call one of my agents to escort your mother down because you are both teenagers and need the company of an adult," the agent replied.

"Thank you," Janice said.

Agent Zorokowski inquired, "So, where is she?"

Janice replied, "She is waiting on 20th and Lemon. I will call her and let her know."

Janice called her mother to inform her that she and J.C. could not come down and that an ATF agent would instead.

"Agent Zorokowski looked at Janice and said, "Can I ask you a few questions?"

Janice looked at him and replied, "Are you serious? I am not in the mood for any questions."

They walked back toward Aunt Kay's home and entered. At that time, Janice and J.C. took a seat. Agents Zorokowski removed a small writing tablet from his back pocket and as he looked directly at Janice.

Clearing his throat, he said, "Okay, first of all, a team of my agents found a detonator used in numerous bombings. We do not understand why this detonator would be used here in Long Beach on a woman with no criminal record or ties to a corrupt organization," Zorokowski explained.

Janice looked shocked by this bit of information.

Agent Zorokowski continued, "Janice, have you noticed anything different about your aunt?"

"No," Janice replied.

Agent Zorokowski further explained, "Okay, sorry, I must ask these questions."

Janice replied, "Well, all I can tell is that I was awakened by a loud boom and the smell of smoke. Then within a couple of minutes, my aunt's neighbor was banging on the door, saying something terrible had happened to my aunt!"

Agent Zorokowski said, "Okay, thank you for whatever information you can provide or recall. Also, can you direct me to the neighbor that came to your door?"

Janice replied, "His name is Smitty."

Agent Zorokowski continued, "Whoever did this is very experienced in explosives. They put a device called a *variable* on your aunt's car."

Janice looked at Zorokowski with confusion.

Zorokowski understood Janice's confusion and replied, "Let me explain it. A variable resistor is a device that controls the flow of electricity in electrical circuits. And this is exactly what that culprit has done. They installed it behind the radio. Anyone who turned it on wouldn't survive."

Meanwhile, Zorokowski's phone rang inside his pocket. As he answered the call, he heard, "Hello, agent Zorokowski, we have the victim's sister," another agent said.

"Okay, we are waiting." Zorokowski's replied.

Agent Zorokowski told Janice, "Your mother is on the way with two of my agents."

Zorokowski turned around, looking out the window; he noticed a couple of firefighters cleaning up. Zorokowski said he could not understand why anyone would take the time to install such a device in Aunt Kay's car. A single mother who worked 40-plus hours a week. She had just been an average woman. Zorokowski noticed a small woman walking up the front walkway ahead of his agents. This woman stood approximately five feet one inch tall, if not shorter.

It was Janice's mother. As the door opened, Janice and J.C. immediately made their way to her. They embraced her, crying into her shoulder.

Zorokowsi met Janice's mother and said, "I am sorry for your loss."

Zorokowski made his way out of the house, but another agent called his name before entering the street.

Agent Zorokowski turned around and said, "What's up?"

"Sir, we have found this metal box!"

The agent handed the box over to agent Zorokowski. He opened it. There were ten oval-shaped diamonds inside.

Agent Zorokowski exclaimed, "Wow! I have never seen diamonds this big before. Who else have you talked about this?"

"Just you, Sir," the agent said.

"Okay," agent Zorokowski replied as he closed the box and tucked it under his left arm.

"What had Kimberly Kemp been up to with all these diamonds?" He mumbled. "Was she living a double life?"

While lost in his thoughts, he noticed an older gentleman crying his lungs out.

Agent Zorokowski said, "Excuse me, Sir, my name is Agent Zorokowski, and I am the lead investigator on this case. Can I ask you a few questions?"

As the older gentleman tried to control his tears, he replied, "Yes!" His voice trembled.

Agent Zorokowski said, "So, how long have you known Kimberly Kemp?"

"Well, I've known her for a long time, something over 20 years." With uncontrollable tears coming down his face, he replied.

Agent Zorokowski inquired, "Did you notice anything different or strange in the past few days?"

"No, Sir. Not at all. She had the same routine. Everyday. Her thing was going to work, you know. Just taking care of her business always."

"What's your name?" Agent Zorokowski interrogated.

"Well, they call me Smitty around here; I've lived in Long Beach forever, in the same house."

Agent Zorokowski smiled at that. He further inquired, "So, how did this happen? I know you keep a keen eye on everything on your street."

Smitty replied, "Yes, Sir. I do. But I have no clue how and why this happened."

Agent Zorokowski said, "If you have any information, you can reach me here."

He extended a business card and walked away. Agent Zorokowski heard Smitty laughing.

"What is so funny?" He asked.

While laughing and holding his stomach, Smitty said, "nothing."

Zorokowski headed toward his vehicle, still deeply thinking about this lady being murdered. Putting his key in the ignition, he drove off with the black box in his passenger seat.

Janice and her mother sat down, looking into each other's eyes. Janice's mother asked, "Are you okay, sweetheart?"

Janice replied, "Yes, I am fine."

J.C. still had not spoken; he just stared at the clock.

"Why?" Janice continued.

Her mother responded, "I know, baby. I am scared for your cousin, too."

They began to cry.

"It's hard to see your loved ones in terrible conditions. I am so sorry for you both having to experience that horrible site."

Janice's mother cried with Janice and reached out to hug her and embrace her to ensure everything would be all right.

Chapter 2: The Beginning of It All

"Quick, please hurry! I have to see my parole officer at 10 A.M., then drop you off at your grandfather's house, and I even have a computer class to teach. I can't be late!" Akbar yelled in a hasty tone.

He taught computer classes at the community Teen Center he opened after being released from prison. Akbar's manslaughter conviction had been overturned on appeal. So, during his time in prison, he took numerous computer classes and taught himself computer coding and programming.

Akbar wanted to make a change in his life and turn things around, knowing that Quick needed him and that without his life skills, he would be of no use to his own son. He wanted to be that positive mentor in Quick's life, just like his father was to him.

Unfortunately, Akbar had made some bad choices in life. He ended up getting arrested during a heist in which computer chips were stolen, and, in the process, the getaway driver was killed.

Akbar was acquitted of the murder but convicted of manslaughter. A few years later, he was released on an appeal. He was okay with it as long as he could take care of Quick, the only person that mattered in his life.

Akbar's father, Mr. Gibson, opened his first black-owned grocery store in the community many years ago. Even with him going to prison, Mr. Gibson remained loyal to his son; he understood that Akbar was a very intelligent man.

Quick's mother left just a few months after Quick's birth, she was an incredibly beautiful Italian woman, but her family never approved of Akbar. So, Akbar stepped up and took the responsibility of raising his child.

Naturally, Quick would often inquire about his mother, but Akbar had erased her from his life and discarded every picture that pertained to Quick's mother.

Akbar inquired, “Are you ready? Quick.”

Quick replied, “I am coming, old man!”

Akbar responded, “I got your old man! Cause you sure weren't saying that when I beat you five times to your zero in basketball!”

With nothing to say, Quick decided not to respond; he grabbed his baseball glove and made his way downstairs. Quick yelled out while grabbing an orange from the fruit basket placed on the dinner table.

Akbar teased Quick, “You know. I let you win.”

He tossed Akbar the orange that he had grabbed off the table. As Quick and Akbar walked out of their home, Akbar pressed the little button on his key fob, deactivating the alarm on his Dodge Ram 1500. It was black on black with a limo tint; Quick loved that truck, especially when he was allowed to pick up Janice with it.

As soon as Quick entered the truck, he reached straight for the remote control that controlled the sound system. But Akbar grabbed it first.

“Oops!” He said, laughing out loud, “Who's the old man now?”

Quick and Akbar both laughed. Quick loved his dad. He was his superhero, an ex-con pops, who became one of the best computer programmers in the world, and he even became his own boss by opening up his own software company. He is proof of the misconception in society that all ex-cons return to prison, which isn't true. Quick just smiled when he thought about all the obstacles his father had gone through.

Abkar wanted to build his legacy, showing his dad, Mr. Gibson, that he could do it on his own. Akbar’s father wanted him to work at their family-owned grocery store. But Akbar had different plans. Akbar started up his truck, and the sounds of Anita Baker blared out of the truck, vibrating the trucks, windows, and seats. The song playing on the radio was *"You bring me joy."* Akbar put his truck in reverse and pulled out of the driveway. He headed toward Quicks grandfather's house.

Akbar looked over at Quick and said, "I hope you have those math questions?"

Quick replied, "Do you mean my homework?" Quick looked over and smirked at Akbar continuing, "Yes, I have them."

Akbar said in a taunting tone, "Good! Because last time I helped you out, you forgot them at home."

As they drove up the neighborhood, Ms. Taylor was out walking her German shepherd. She smiled and waved; her huge smile illuminated her face. Akbar smiled and waved back, as he knew Ms. Taylor liked him because she had made numerous moves on him in the past.

In fact, Quick was also really close friends with her son. Quick looked over at his father and joked, "I know you want to hit that!"

They both had a good laugh over it.

Quick further added, "Yeah, knock those boots!" Quick laughed hysterically.

Laughing, Akbar responded, "Hold on. We will get back to that. I need to call your grandfather."

Akbar pressed his onboard contacts, but before the call could be connected, he noticed an unmarked vehicle trailing behind him. It was about two cars' lengths behind him; he could see it in his rearview mirror.

Mr. Gibson answered Akbar's call and said, "Hello, Akbar."

Akbar said, "Hey, dad, I'm on my way."

At the same time, he was paying close attention to that unmarked vehicle behind him.

"Okay!" Mr. Gibson replied.

They hung up the call, and in less than two seconds, a patrol car appeared and flashed its lights; it pulled up directly behind Akbar's truck.

He yelled, "Damn it. What the F*CK!"

Akbar pulled into the strip mall, having found a safe place to park.

Akbar instructed Quick, "Stay calm, Quick. I do not understand what's going on, son."

As he put his truck in park, he looked at his side-view mirrors and rearview mirror. He noticed two officers walking toward his truck—one on Quick's side and one on Akbar's.

One of the officers tapped on Akbar's window.

Officer number one asked, "Can you please exit the vehicle?"

You could see intense anger on Quick's face; he would call it the harassment bucket, meaning all African Americans being put in the same category. One bucket. Suddenly, there was a tap on Quick's window as well.

It was officer number two. He said, "Excuse me, can you please step out of the vehicle?"

All this time, officers kept their hands on their service pistols. As Quick opened his door slowly and stepped out, the officers could not believe how muscular he was.

Being only 15 years old, Quick stood six feet four inches tall and roughly weighed two hundred pounds. The officers seemed very intimidated by his size.

Officer number two gave orders for Quick to sit down in the car. Quick so desperately wanted to say fuck you! But he had to control his emotions, something his grandfather had taught him.

Akbar walked toward officer number one's patrol car, and he turned around, asking them what exactly was going on. But to his surprise, without ordering, officer number one suddenly grabbed Akbar by his waist and immediately handcuffed him. As if it wasn't enough, officer number one handcuffed Quick as well.

Akbar screamed, "Why are you putting my son in handcuffs?"

Officer number one replied, "I'll let you know in a minute."

Officer number two walked Quick to an unmarked vehicle.

Enraged, Akbar shouted, "What the fuck? You do not have to handcuff my son and put him in the back seat. He is no criminal."

Officer number one furiously said, "Shut the F*CK up!"

Then he grabbed Akbar by his arm and walked him toward a black van with limo tint. He could not believe what was going on, a patrol car pulling him over, his son being handcuffed, and two unmarked vehicles, as well as a van and a Lincoln town car.

Straining his eyes, Akbar tried to look inside the van as he walked toward it, but the windows were too dark. You could hear the vehicle motor; it was still idling, ready to take off when needed. Looking around, Akbar tried to read the names on the officer's badges, but things were moving so fast he could not make them out.

Snap!

The latch was undone on the side door of the van, and it opened. A tall man exited the driver's side of the van. He walked toward Akbar and politely asked him to please step in. This man wore all brown clothing, with black leather gloves. Akbar was directed to have a seat in the rear of the van. As he stepped into the van and took a seat, the door closed slowly.

Unexpectedly, a deep male voice came through a set of small speakers in all four corners of the van and said, "Hello, Akbar! Or should I call you Cleavon Gibson?"

Akbar knew whoever it was had done their background investigation on him.

The anonymous voice continued, "Mr. Gibson, what I want to know is where those computer chips are and to whom did you sell them? Oh, and just to let you know, somebody sold you out! That pallet was worth only millions, but it is not the monetary value, but rather what they are capable of doing. That's what makes them valuable! So, what I am saying is if they fall into the wrong hands, this country might find itself in turmoil. China and Russia are constantly hacking our country's infrastructure. Those computer chips were made specifically for artificial intelligence. Android robots that look so real that they even have their own identification print. Meaning their own fingerprints. So, it is important that we get those chips back!"

Moved, confused, and scared, Akbar said, stammering, "I do not have any information for you. I do not even know what you are talking about."

The anonymous voice replied, "Well, just to let you know, whoever has them, you better hope they do not have terroristic tendencies. They could be disastrous for all of us."

A little settled now, Akbar replied, "Like I said, I have nothing for you, so can you please let me and my son go?"

The voice replied, "Sure, I can do that."

Akbar was separated by a partition between him and the voice from a small speaker just above him.

The voice said to Akbar, "We have been watching you very closely."

Suddenly, the van door opened.

Akbar looked over and noticed the officers staring at him. It gave Akbar the chills. Officer number one said step out, as Akbar hurried toward the door to exit the van, he was instructed to turn around. The officer then uncuffed him and pointed toward the unmarked car. Officer number two opened the vehicle's door, releasing Quick.

"Something is going to change in your life," Officer number one whispered to Akbar, leaning into his right ear and smirking as he walked away.

Akbar looked completely confused about what was going on, but whatever it was. He would eventually figure it out. But at that moment, he just wanted to get Quick far away from the scene. Akbar stared at Quick, knowing he had to keep a close eye on his son.

He thought, *who could have sent this team? I have been out of prison for a while, and I have started my own company as well. Could it be China or Russia? I do not know.*

But this team is English-speaking; something just doesn't add up. Nothing makes sense. Law enforcement? Akbar thought. He just shook his head in wonderment.

As Quick opened his passenger door and sat down, Akbar and Quick looked around, watching the officers as they returned to their vehicles. Quick looked incredibly angry, knowing whatever his dad had done did not sit well with them.

Suddenly, Quick realized that one of the officers was named Alvarez; Alvarez was part of a gang task force. He remembered him because he and two of his friends had been harassed by that officer in the past. It did not make sense why Alvarez would be part of this.

"Damn!" Akbar yelled out.

Akbar dialed his dad's number, and after a few rings, he answered the call and said, "Hello, are you close?

Akbar replied, "No, Quick, and I got pulled over. So, we are running late."

Mr. Gibson said, "Pulled over for what?"

Akbar replied, “No reason, just being harassed.”

He asked his father to hold on as he switched to another call coming in.

Akbar said, “Hello, Laura. Is everything okay?”

Laura answered, “Not good, Akbar, not good at all. I just got a call that Jimmy has been shot with a bullet wound to the head.”

Unable to comprehend what was going on, Akbar cried, “What?”

Laura could be heard crying on the other end of the phone.

Laura continued, “And that's not all. His throat was sliced open from ear to ear.”

Akbar gazed out into the distance as he could not believe what he heard over the phone, and neither did he have any words to console Laura back.

The last thing Akbar wanted to hear was that Jimmy was dead. Jimmy was a close and good friend to Akbar, and he introduced Laura to Jimmy. Akbar and Laura had been friends since childhood. Laura graduated from Stanford, and they finished law school together. She was the one that had helped Akbar with his appeal. Even though she was a corporate lawyer, she represented Akbar whenever he needed her services. Laura's body of work was impressive. She was one of the top ten corporate lawyers in the country.

Akbar told Laura, “I will be there in a few hours.”

Laura replied, “Okay, Akbar, I will be waiting.”

Akbar hung up the call and returned to his dad, who was still on hold.

Akbar replied, “Pops. I am on my way; something has happened with Laura. I will explain it to you later.”

Akbar put his Dodge Ram in sports mode, looked in his mirrors, and raced right into traffic. He hit his playlist and played the “*Lyfe Jennings*” album. The song “*Keep Your Head Up*” blared through this truck radio. Akbar couldn't believe how the morning was turning out. He even started thinking about how he or Quick could possibly have been killed. Even the thought of him going back to prison crossed his mind. He could not imagine being away from Quick again. The thought made him sick.

As Akbar approached the 710 freeway, he accelerated his Dodge Ram, reaching one hundred mph. What a change of events in one short morning. Getting pulled over, then hearing what happened to Jimmy. Wow! These thoughts kept racing in his head as he changed lanes, going up to the 710 freeway. They suddenly approached Willow Ave and exited the freeway.

Akbar called his dad and said, “Hello, Dad. I should be there in a few minutes.”

Mr. Gibson replied, “Okay, son.”

They both hung up the call.

Akbar looked over at Quick and said, “I will be a little late picking you up after baseball practice, I will call if I am going to be late, and if so, coach Rogers can drop you off at Dad's house.”

Quick responded, “Okay, but is everything all right?

Akbar replied, “Yes, things are fine. I just need to make a few stops.”

Akbar knew that heist had awakened a sleeping giant regarding these computer chips. They could change the course of the world, knowing that those chips were supposed to be used for military artificial intelligence robots. And if they fall into the wrong hands, things could turn out all bad for society.

As Akbar drove to his father's home, his nose started bleeding.

He said, “Oh shit! What a fucking day.”

His nose would bleed whenever he began to overthink.

Quick said, “Here you go, dad,” handing him some tissues.

Mr. Gibson was standing in the front of his home, awaiting Akbar, pulling up close to the curb. He rolled down his passenger window.

Akbar said, "Hey, dad, here is your twin."

Mr. Gibson replied, "You made it fast. I was not expecting you for another 15 minutes."

Akbar replied, "Yeah. Well, I push the pedal to the metal, pops."

Quick exited the vehicle; looking back, he said, "Do not forget to call me if you are running late, dad."

Akbar gave Quick a thumbs-up and rolled up the passenger window while driving off. When he approached a stop sign, he placed his hand under his seat, reaching for a bottle of vodka that he downed in a single gulp. Shaking his head, Akbar made his way back toward the 710 freeway. He needed to pay Skip a visit to find out what he knew.

Skip was the one to put this heist together, but Akbar fine-tuned it. Skip worked for the Port of Los Angeles and held a position as a customs agent. So, if anybody would know, it would be Skip. He was a very shady customs agent and had been under investigation once or twice before.

Beep…beep…beep, a call coming through Akbar's phone. It was Laura.

Akbar said, "Hello Laura, what is up?"

Laura replied, "Well, just to let you know, the detectives are at my office. Just wanted to let you know."

Akbar said, "Thank you, Laura, but I need to stop by Skips first."

Laura replied, "Okay."

As Akbar drove toward the 710 Freeway Northbound, he couldn't believe how they knew who he was and how they found him. Those diamonds were never mentioned in the robbery.

Skip owned a 107 feet Vince cruiser yacht that docked in Marina del Rey. It was custom-built in a Turkish shipyard. People always questioned how a customs agent could afford to live on a yacht at Marina del Rey.

Skip's parents had left him several properties throughout California and even out of state. He was awfully bad at managing money. Skip and his sister always fought about money, and his temper was terrible.

One time, Skip invited Akbar to a Sade concert in which Skip punched one of the security guards because he was not allowed backstage.

"Hello," one of the women waved at Akbar.

Akbar politely waved back as he pressed the keypad code to enter the Marina. Ms. Robinson was standing nearby.

"Hello, Ms. Robinson. How have you been??"

Ms. Robinson replied, "I have been doing just fine."

Akbar smiled, "That's wonderful." He paused, "Hey, Ms. Robinson. Just a quick question, have you seen Skip?"

"No. I have not seen Skip for two days."

"Okay. I did not see his car out front." He walked aboard the Vince yacht and looked around.

Something felt out of place as he walked toward the rear of the yacht. He noticed blood on the sliding glass door. As he stood and stared at the blood, he got cautious before entering, thinking about where the blood came from.

Ms. Robinson had not seen Skip for several days; it was not something Skip would do. He wouldn't vanish and not check in on his pride and joy, this big, massive yacht. That is not like Skip at all, Akbar thought to himself.

"Hey, Skip?" He called out his name several times, yet there was no answer.

Akbar made his way toward Skip's bedroom. The massive yacht had a dance floor and a wet bar in the game room. Akbar hesitated before entering the room, especially after he noticed the blood on the sliding glass door. As he entered Skip's master bedroom, he immediately saw that the sheets were soaked with blood.

“What the FUCK!”

He pulled the sheets back slowly and was shocked as he noticed Reva laying in a pool of blood. What stood out is that someone had carved the word sheepskin on her chest as she lay naked with one bullet in her head. The killer had shaved all her hair off.

Akbar stood back, wondering if he would find Skip in the same condition. He walked out of the master bedroom, calling for Skip, but there was still no answer. Akbar was worried that he might find Skip the same way as Reva.

The yacht was huge, with so many places where Skip could hide. Akbar searched the entire yacht, but there was no sign of Skip anywhere. He knew he had to clean up before exiting the yacht. Before returning to Riva, Akbar retrieved a body bag that Skip had kept on board. They were used when they went deep ship PC fishing for Marlins.

Returning to Skip’s master bedroom after he placed the bag beside Skip’s California King size bed, Akbar held Reva by her wrist and pulled her from the bed, and put her body into the bag. It was a clear bag filled with blood from Riva’s lifeless body. Akbar zipped her inside the bag and removed all the linens from the bed. Even the mattress was soaked in blood. Akbar would have to come back for that later.

He dragged Riva’s body down to the walk-in freezer and placed her inside. Akbar then turned back toward Skip’s room, grabbing Riva’s clothes and linens and putting them inside a giant trash bag. But before exiting, Akbar removed his clothes, searched through Skip’s closet, and took some of Skip’s clothing. But before he could get dressed, he looked out the window and noticed a drone hovering outside the window.

“What the FUCK?” Akbar yelled and dropped to the floor, carefully peeping out the window, not knowing what was happening.

Why would a drone be outside this window? Akbar grabbed everything as he got dressed out of the drone's view and peeped out the window once more when he noticed the drone had disappeared. He jumped up, moving and grabbing the items he had put in the bag, and exited the yacht with the bag.

He then walked toward his vehicle, looking around to ensure he was not being followed. He could not help but think of all of the events that were going on.

In such a short time, Laura's boyfriend Jimmy was dead. Akbar and Quick had been pulled over, Skip was nowhere to be found, and now Reva was dead. All of this was way before 10 A.M. He wondered why people around him were being killed, People who had nothing to do with the heist. It was just a confirmation that someone was pissed off.

Akbar recalled what the officer had said earlier that morning, how things would change, and that they were changing too fast. Akbar's mind replayed the word *sheepskin* carved in Riva's chest and a bullet hole in her forehead.

So extreme! Putting a bullet in her head.

Akbar thought about Quick and his father. His phone vibrated, and Akbar almost dropped the bag with all the bloodstained linen. Looking down at the scene, an unknown number appeared.

Akbar touched the screen, and a voice on the other end said, "Sheepskin."

It was a male voice. "I have a package for you at the Holiday Inn on Figueroa. Report to the front desk. You are registered under James Campbell. When you receive the key card, proceed to the room number on the key card, and please wait for my call.

Also, I would not come back for the young woman. A call was placed to report a suspicious person was on board that yacht. So, I advise you to move on it. The sheriffs are already on their way."

The call dropped. Akbar looked around and pressed his key fob to unlock and start his truck automatically. But before getting back into his car, Akbar scanned the area, knowing that someone had been watching all his moves. First, the drone, then the unknown number; the strangest thing was that they had his phone number.

Akbar climbed into his Dodge Ram pickup, put his truck in reverse, and exited the Marina. He engaged his onboard voice recognition and dialed Quick's phone number.

"Hey, son. I know you are in school, but most likely coach will have to drop you off."

Quick replied, "Okay!"

"You might have to stay over for the night."

"That is fine by me. Janice and I can watch a movie later tonight." Quick replied.

"Okay, sounds good, son. I love you!"

Akbar hung up as he approached the 405 freeway. There was still a bit of morning traffic, which was normal for California. Akbar put on smooth jazz, some of Kenny G's greatest hits. He headed toward the 110 freeway, trying to relax to those smooth melodies as he thought about Reva and her lifeless body.

So many things ran through Akbar's mind, but what stood out was the strange call he received, and now he was en route. Akbar knew he had no choice but to go or end up attempting to explain what happened to Reva once the sheriff arrived.

Paying attention to slow traffic, he began to accelerate at normal speed, replaying in his mind repeatedly all that had transpired in the morning hours; How Ms. Robinson said she had not seen Skip for two days—which was unlike him.

Akbar thought that he still had access to the remainder of the computer chips and that ten million had already been given upfront for them. Being in such deep thought, Akbar never noticed that he missed the exit.

His 30-minute drive came by fast as he exited. Akbar took another shot of vodka. Three blocks away, Akbar noticed the Holiday Inn. He accelerated his truck, and you could hear the exhaust as he pulled into the Holiday Inn parking lot.

Before pulling into the parking spot, Akbar looked around. After he felt comfortable, He parked. Everyone seemed like a suspect to him. He noticed a young black man standing near the entrance. Akbar walked into the Holiday Inn and approached the front desk. There stood a young Indian woman typing on her computer.

As he stood there waiting, he noticed how beautiful she was. She had jet-black hair and a chestnut-brown complexion. She looked up and smiled.

What a radiant smile she had!

The Indian woman said, "How may I help you, Sir?"

Akbar replied, "Yes, my name is James Campbell, and I am here to pick up my room key."

"One moment, please," She replied with a smile. She put his name into the computer and located it. "Okay, sir. You're booked in room 2011, located on the second floor."

She then handed Akbar his room key, looking directly into his eyes.

"Thank you!" Akbar nodded at her and walked away.

He read her nametag and memorized her name, Sandra. He would not be quick to forget it either. Akbar walked through the hotel lobby and toward the elevator. As the elevator door opened, Akbar pressed the number two.

Once the door closed, it seemed as though they reached the second floor at lightning speed. Once on the second floor, Akbar exited cautiously, looking around in both directions. Akbar's heart began to beat faster and harder, not knowing what to expect, especially after the call.

Akbar noticed that room 2011 would be toward the left as a room indicator on the wall pointed to the left for rooms 2000 through 2020. Akbar looked back and forth for a while as he continued to walk toward room 2011. As he reached the room, he inserted his keycard. The hotel door unlocked, and Akbar entered the room. He could hear the television in the room was on.

Once inside, Akbar could hear the national Business Report with Robert Risca was on. Akbar scanned the room and noticed a cell phone placed on the bed. Suddenly the phone rang. Akbar walked slowly toward the phone. He reached down and answered it.

"Hello?"

A male voice came through and said, "I am Dr. Zu Cheng Zhang. I am a neuroscientist at the University of Wisconsin."

Akbar replied, "It is nice to meet you."

The Dr. immediately started telling Akbar about some investments he had made that had turned out terrible for him. He was referring to the military artificial intelligence soldiers. It was at that moment when Akbar had to interrupt him.

"Excuse me, Sir. I have no idea what you are talking about."

"I do not believe you," the Dr. replied, and the phone call went silent. The Dr. suddenly yelled out, "Operation sheepskin!"

The phone call immediately dropped, and Akbar could not believe he had come all this way to hear someone ramble on about things he knew nothing about. What weighed heavy on Akbar was the word sheepskin. He knew that Quick and his father were also in danger.

What have I done?

Akbar understood that computer chips were for a special military project which did involve making artificial intelligence soldiers. At this point, Akbar was all in. There was no going back. Two people had already been killed, and Skip could not be reached. Akbar broke the phone into pieces and flushed it down the toilet.

He then exited the room, looking both ways. That is when he immediately noticed an Asian man standing to his right. Akbar reached into his pocket for his phone and dialed a close friend Vic. The smooth-sounding voice on the other end answered.

"Hey, Vic. Tonight's meeting will be early. We must put a rush on it. I cannot explain now. I will get back to you later." Akbar spoke in a hurry and disconnected the call. He could not stop thinking that those computer chips had a lot to do with Reva and Jimmy being killed. Skip must have gotten cold feet. Akbar had told Skip about the last sale of those chips; they should have gotten rid of them all back then. This entire heist had become toxic.

Akbar checked the time on his watch and said loudly, "Damn!"

He realized he was running late to see his parole officer, who hated him. The fact that Akbar became successful in proving them all wrong made the parole officer dislike him even more, so he hated when he ran late. Akbar turned back toward the hotel clerk's desk before leaving the hotel. He handed back the key to Sandra and asked if she was free for dinner.

Sandra replied, "Sure."

Akbar smiled and gave Sandra his number on the Holiday Inn business card.

He asked, “What time shall I pick you up?”

“Not tonight, but do not worry, I'll call you,” Sandra replied as she took his business card.

Chapter 3: Fostering a Relationship

Mr. Gibson, Akbar, and Quick would have family meetings weekly. Akbar pulled in front of the lighthouse restaurant located in downtown Long Beach. The three of them exited the vehicle and walked toward the lighthouse restaurant.

Gearing up for their family meeting, they would call it a brain teaser weekend. Their loyalty to one another was amazing. Akbar just wanted to make sure that Quick did not make the same mistake he had made in his youth - that he did not fall victim to the regular circumstances of someone like him. Akbar wanted to ensure that Quick maintained his education and, secondly, learned about his history.

Quick was a very well-mannered young man; he and Janice were junior high school sweethearts. She meant everything to Quick. Akbar noticed a lot of similarities between Janice and Quick.

Two years had passed since Akbar had seen Skip or Vic. By this time, Akbar had established himself in the computer world. As a consultant and engineer. He also dealt in contact with government agencies. He knew they were in love because he loved Quick's mother, the same way Quick loved Janice. As they entered the Lighthouse restaurant, their favorite waitress, Lucy, greeted them.

Lucy smiled and said, "Hello, gentlemen."

Mr. Gibson responded, "Hey Lucy, how have you been?"

Lucy replied, "Well, I have been just fine. Mr. Gibson, just fine!"

The Lighthouse was famous for its hot wings and had some of the best catfish on the West Coast. The Los Angeles Times had ranked them number two out of the top ten places to dine by Anthony Costello, who would write restaurant reviews. Lucy waved the gentleman over.

Lucy instructed, “Follow me this way, please.”

And the three were taken to their usual seating. A family who had been waiting for 30 minutes stared and just looked at them as they were seated. Two women also pointed at them—in shock, not understanding why they were taken first.

As they approached the table, Quick reached for his phone and called Janice.

After a few rings, Janice answered, “Hello!”

Quick responded, “Hello, my love, we are at the lighthouse, and I just wanted to let you know that I will be bringing back some shrimp tonight.”

Janice said, “OK, thank you, my baby!”

Quick responded, “I have to go, baby; I will talk with you later....”

They both hung up the call.

Quick knew that Janice would be hungry after the karate class. Lucy directed the three men to their seats. She handed them a menu, even though she knew what they would most likely want to order.

Each week someone would pick a subject to speak on. Mr. Gibson would say what is going on, lover boy? Quick just continued to smile. Every time he spoke with Janice, Quick's face would shine. He felt the burdens of life had been lifted off his shoulders.

Quick removed his glasses, a brand-new pair of Calvin Klein's that he had gotten for his birthday. Quick was very mature for his age. He had two extraordinarily strong male figures to support him: his grandfather and his father.

Mr. Gibson had even run a few marathons for charity. He was a highly active member and leader in the community. He ran a well-known not-for-profit organization for kids whose parents were incarcerated. Mr. Gibson opened one of the first black-owned grocery stores in the community. Mr. Gibson's wife had died during surgery due to negligence by Dr. Raphael De la Roca. The doctor had tested positive for opioids and was found to have been under the influence while performing surgery on Mr. Gibson's wife.

Mr. Gibson looked over at Quick and said, “You must have talked to Janice?”

Quick replied, “Yes, I did. I told her I would be bringing her home some popcorn shrimp after we were done.”

Mr. Gibson responded, “Well, I see we have raised a young gentleman because your father never had swag like you, Quick.

Quick responded, “What do you know about swag?” Quick smiled at his grandpa.

Mr. Gibson said, “I know a lot about swag. Social media teaches you so much these days.”

Lucy returned to take their order, “OK, gentlemen, what will it be? The usual?”

Mr. Gibson said, “No, I will just have a chef salad.”

Lucy added, “Excellent choice, no carbohydrates.” Lucy smiled as she looked toward Quick and Akbar.

Quick said, “I will take the chicken strips.”

Akbar asked her the same.

Quick requested Lucy, “Can you put popcorn shrimp on that order as well?”

Lucy replied, “I sure can!”

Quick said, “Thank you, Lucy!”

Mr. Gibson started their meeting by saying they would discuss the first encounter in the West African societies by the European trade.

"Sure, sounds good," Akbar replied.

Mr. Gibson said, "Well, during the era of the European slave trade, which was roughly from the 15th century through the 19th century. Numerous enduring myths about Sub-Saharan West Africa were propagated. Even today, they claim that the people who inhabited this region for hundreds of years were isolated from the rest of the world. They had simple, self-sufficient economies. Some scholars still depict the region stretching from the Senegal River South to modern Angola. As a single culture unit. As if one time all the women and men living there must have shared a common set of African political, religious, and social values."

Quick and Akbar continued to listen while Mr. Gibson took a break and paused.

Quick said, "My teacher talked about the Portuguese who explored the African coast."

Mr. Gibson said, "Yes, that is correct. They explored around or during the 15th century. The Portuguese explored a vast variety of political and religious cultures."

Lucy returned with their orders.

Lucy said, "Catfish, chef salad."

Lucy already knew Mr. Gibson wanted the catfish, just without the potatoes, and substituted it with a chef salad.

Lucy then said, "And two orders of chicken strips. Quick. I will return with the popcorn shrimp before you guys leave."

Right before Akbar could take a bite, his phone rang. Akbar said hello; the voice that came through said, sorry for the interruption, boss. There has been a break-in at the office. You need to get here immediately before I call the police. Akbar told him that he would be there in 30 minutes.

After Akbar hung up the phone, he said, “Fellas, we need to wrap this up. There has been a break-in at the company.”

Mr. Gibson said, “What?”

Akbar explained well, “Let's get a doggy bag.”

Akbar waved Lucy over, and she walked toward the three of them.

Lucy said, “Well, are you already done?”

Akbar responded, “No, Lucy, we have an emergency, and we have to go.”

Lucy said, “OK, let me get those popcorn shrimp and three containers for your meals.”

Lucy walked away, and Akbar went into deep thought, wondering who would break into his Huntington Beach office. After a few minutes, Lucy returned with three Styrofoam containers and the popcorn shrimp. Akbar reached into his back pocket, pulling his wallet out.

Akbar said, “Here you go, Lucy, thank you.”

He handed her a $100 bill and told her to keep the change. As they exited the restaurant, Akbar called Seyne.

After three rings, he finally answered, “Hello, boss.”

Akbar instructed him to look around the office and tell him if something was missing or destroyed. The call suddenly dropped, so he could not get the information from Seyne. Akbar drove up the highway, not even paying attention to his speedometer. He was still uncomfortable; He could never truly relax. Akbar was aware that the millions he had acquired in that heist were blood money. Even after going legit, he could not shake that idea out of his mind.

After a brief drive, Akbar pulled into his parking spot. His parking spot read, “Akbar D. Gibson, Owner, CEO.” Akbar turned off his car, looked over at Quick and his father, and gave them orders to stay inside the vehicle until he returned. Akbar walked toward the entrance. He put his thumbprint on a black security pad that allowed him access to his company. "*Beep, beep,*" the massive glass doors opened, and Akbar stepped in, looking around and observing the building. It had well-maintained floors where you could see your reflection. It was as if you were looking in a mirror. As he approached the elevator doors, he pressed the button for the third floor.

"*Ding,*" the elevator doors opened.

Seyne said, "Hello, boss!"

Akbar stepped out of the elevator and walked toward Seyne. Akbar knew something was seriously wrong from the look in his eyes; the look said it all.

"Boss, there is something you need to see. Please follow me."

Seyne pointed toward the employee's bathroom. Both men walked toward the Men's restroom. Akbar entered with caution, and once inside, he noticed a red fluorescent light had been installed, replacing the regular light bulb in the restroom. So many things had been going through Akbar's mind by this point.

Please, no, not another dead body, he thought.

Akbar noticed a large trash bag sitting in the middle of the bathroom floor. Akbar called for Seyne and asked him if he had looked inside the bag. Seyne responded, saying that he had not and was waiting for Akbar to reach first. Both Akbar and Seyne approached the bag with caution. They knelt beside the bag and opened the bag slowly. Akbar grabbed the bag, looked inside, and immediately dropped it. It was a sheepshead. Akbar punched one of the glass mirrors inside the employee bathroom. All he remembered was Reva and the word sheepskin engraved on her chest.

Seyne yelled, "Boss, boss. Your hand. It is bleeding badly!"

Akbar grabbed the bag and walked back to his office.

Seyne asked, “Should I call the police now?”

Akbar answered, “Yes, now you can.”

Seyne called the police department. Akbar then opened his office door. As he walked toward his desk, he noticed a set of photos on the desk. There were photos of Quick's mother. He wondered how and why these photos were there. He dropped the photos on the desk and gazed out his office window. Someone was sending him a sign. Akbar heard a voice, and he looked up and noticed it was Quick.

Quick was making his way into Akbar's office, so Akbar Quickly stuffed the pictures in his back pocket. He would not show the photos to Quick, as he knew it would break his heart if he ever found out his mother had just walked out of his life.

Akbar thundered, “Did I not tell you both to wait in the car?”

Mr. Gibson replied, “Fuck, we got tired of waiting!”

Quick commented, “Wow, it seems like they destroyed the place.”

Akbar replied, “Yes, they did, but it could have been worse, and they did not take anything. What they did do is leave a severed sheep head inside of an employee bathroom.”

Mr. Gibson replied, “Why would anyone think of doing something sick like that?”

Mr. Gibson was upset once he started cursing.

Akbar ordered Seyne, “Please send everyone a text message and inform them to take the day off!”

Seyne replied, “OK! I will do that immediately, and I will start cleaning up.”

Akbar took a seat as he thought about those photos he found in his office. He knew that one day he would have to tell Quick why his mother had departed from his life.

Quick said, “Hey, dad. Are you all right?”

Akbar replied, “Yes, I am fine, Quick, but there is something I would like to tell you.”

Akbar placed two photos on his desk and looked at Quick. Quick slowly walked toward Akbar and asked him what they were. Teardrops started to fall from Akbar's eyes.

Akbar answered, “That is your mother.”

Quick, looking closely without picking up the photos, said, “Why are you showing me this now?”

Akbar said, “I do not know, son. I found them laying on my desk.”

Quick looked confused because he had never seen his mother.

Akbar said, “Son, I just wanted to explain a few things, I know it is not the right timing, but it is perfect for me. Quick, your mother loved you so much, and I still know that she genuinely cares for you, but her family put so much pressure on her when they found out she had conceived my baby.”

With tears rolling down his eyes, Quick said, “So the easiest thing for her to do was leave?”

Akbar uttered, “Yes. I guess, son, I also need to tell you a few more things. Son, I have done some things to advance myself in society further, and it's not all kosher.”

Quick said, “What do you mean? Advance yourself in society?”

Akbar replied quickly, “I am not proud of what I am about to tell you.”

Akbar paused as he thought of the words on how exactly to explain it all. Then he came right out with his nervous ramble and told Quick, “I committed a heist, stole computer chips, and sold them for millions.”

Quick was shocked.

Akbar replied, “Yes, son, I am tired of keeping this from you. The computer chips were made for artificial intelligence soldiers. With a longtime friend, I created an artificial intelligence robot humanoid using these computer chips.”

Quick still could not believe what he was hearing.

Akbar said, “Quick, I am scared that whoever broke in here is sending me a message, and it has to do with that robbery. So, now is a perfect time for me to make you aware of what is happening.”

Quick replied, “Dad, I am glad you are telling me.”

Akbar replied, “Do not think I put you and my father in danger. The soldier I built will now be activated for personal security. Now that my company has been broken into and finding these photos and this sheep head, I have no other choice.”

Akbar called out to Seyne, “Seyne, come here.”

As Seyne approached Akbar's office, Akbar said, “Seyne, can you do me a favor and drop Quick and my dad off at my dad's grocery store?”

Seyne replied, “Yes, sure, boss.”

Akbar stated, “I will clean up later.”

Seyne and Quick walked out of the office. Quick could not take his eyes off the photos Akbar had shown him. He had mixed emotions. First, with Abkar, finally showing him pictures of his mother. Secondly, coming up with what was going on and about the $1,000,000 heist, he built an artificial intelligence soldier to top it off.

Wow! Quick thought to himself.

At the end of the day, he had Abkar and his grandfather in his corner. Quick began to think about Janice and how much she meant to him, how her Aunt Kay had been killed for no known reason.

Quick and Mr. Gibson walked toward the elevator as Seyne escorted them. Akbar called Quick on his phone, but his voicemail picked up.

Akbar said, “Quick, I have a flash drive in my safe that contains five account numbers. If anything ever happens to me, you will always be secure, and Vic will be your most trusted confidant, including the artificial intelligence soldiers. I love you, son!”

Seyne, Mr. Gibson, and Quick reached Seyne's Ford F-150 truck. Quick felt like a burden had been lifted off his shoulders now that he finally knew something about his mother. Seyne started his vehicle and exited the employee parking lot. Quick called Janice to let her know he would be meeting with her later. Quick wanted to know more about the woman in the photo.

Quick said, “Grandpa, can you tell me more about this woman in this photo?”

Quick handed the photos to his grandfather.

Mr. Gibson looked over the two photos and said, “Well, yes, her name is Aurora Girardi, and she was a very beautiful young Italian woman that your father fell in love with. She and your father had been deeply in love until her family stepped in as they disapproved of their relationship with your father. Your mother tried to work things out with your father, but her family made her so uncomfortable that she had to leave. She thought the best thing would be to leave you with us. After she vanished without a trace, your dad tried to find her, but nothing ever came of this search. He did not want to tell you, Quick; he did not want you to suffer, so he destroyed all the evidence of her existence from your life.”

Quick could not believe that the woman that gave birth to him could just pick up everything and leave. But Quick understood that it must have been hard for Akbar. Even when Akbar went to prison, Mr. Gibson took great care of him.

Quick thought about Janice becoming pregnant one day, knowing he would never abandon Janice. Seyne approached Mr. Gibson's grocery store and came at a slow speed; maneuvering through traffic as he pulled up, he put his car in park. Mr. Gibson and Quick exited the vehicle.

Mr. Gibson said, “Thanks for the ride, Seyne.”

Seyne replied, “No problem, Mr. Gibson.”

The two walked toward the entrance of the grocery store. A tall, stocky man approached Mr. Gibson and Quick before they could enter the grocery store. The man put his hand directly in the middle of Mr. Gibson's chest; he was shocked.

Mr. Gibson said, “May I help you?”

The man said, “Yes, you can tell your son we are carefully watching him.”

Mr. Gibson could not help but notice this man had a glass eye. He turned toward Quick and told him to get inside his office. Quick ran inside, not looking back. Roger, a store employee, was busy stocking shelves. Quick moved so fast past Roger that he damn near knocked Roger over. Quick ran to the office and punched in the security access code 615119. The office door opened. Quick immediately walked toward the security monitors in the far-right corner of the office. He observed three men harassing his grandfather. He moved away from the monitors going for help. As Quick exited, he noticed Roger and Stan holding a conversation.

Quick blurted at Roger and Stan, “I need your help!”

Roger said, “Sure, Quick. What's up?” Roger was also a longtime family friend.

Quick informed him, “Well, my grandfather needs help. There are three men out there harassing him!”

Roger and Stan immediately walked out front to help Mr. Gibson. As they approached them, the three men noticed Quick coming back with

Roger and Stan. Both stood six feet tall and 285 pounds. The man with the glasses nodded his head at the other two. They looked around and noticed Quick, Roger and Stan were en route to help Mr. Gibson. The three men with Mr. Gibson departed quickly in an unmarked car that pulled alongside the curb. When the man with the glass eye opened the passenger's front door, he then turned and looked at Mr. Gibson and smiled.

Stan worriedly asked Mr. Gibson, "Are you OK?"

He replied, "Yes, I am fine, son, thank you."

Quick commented, "Today has been nothing but a handful of events. All this in one day!"

Mr. Gibson said, "Tell me about it and the day is still young.! Hopefully, nothing else will happen."

Mr. Gibson requested Roger, "Can you please lock up, and I will pay you overtime?"

Roger said, "No problem, Mr. Gibson. I will do that."

Mr. Gibson told Quick to come with him, and they walked toward Mr. Gibson's 69 Riviera. They both sat in the car in deep thought.

Quick said, "Grandpa can we stop by the gym? I have this popcorn shrimp for Janice."

Mr. Gibson said, "Sure, Quick, call her!"

Mr. Gibson began to exit the parking lot of his store when suddenly, out of nowhere, an unmarked vehicle put on flashing lights and pulled directly in the back of Mr. Gibson and Quick.

Mr. Gibson said, "Damn, now what?"

They were not even a few blocks away from the store. Mr. Gibson became concerned about everything that had happened early in the morning. Quick called Akbar immediately on his emergency number. After a few rings, Akbar answered.

Akbar said, “Hello, son. What is going on?”

Quick replied, “We are getting pulled over by the police. Is something going on? Grandpa had a run-in with three men who stood out front of him at the grocery store, and now three more unmarked vehicles just pulled up.”

Akbar asked where they were located; he told him they were just a few blocks down the street from the grocery store on Seashore Drive.

Akbar said, “OK, I am on my way!”

Akbar hung up the call and headed toward the elevators. Suddenly one of the officers opened Quick's door and pulled him from the car. Quick looked up and thought, *this guy stands six feet seven inches, maybe 280 pounds*. Quick jumped up to his feet, giving one officer a roundhouse kick, and landed his blow to his jaw.

Two more officers who were standing close by took immediate action. The two of them got Quick and pinned him to the ground. Quick looked up at both officers who had done this to him. Quick had broken the officer's jaw, so they put him in handcuffs, restraining him to the ground. Quick's nose was bleeding because the officers slammed his face to the floor. One of the officers tried talking to Quick, but he could only hear a mumbled conversation in the background. Suddenly Quick heard one of the officers yell out.

“Watch it. He has a gun,” and gunfire could be heard immediately after that.

Boom, boom, boom, boom, boom!

It sounded like a Mexican cartel had it out with the Federales: numerous rounds, sixty shots or more riddled inside a nearby unmarked car. Once Quick was inside, he began to scream as he realized his grandfather was slumped over the steering wheel of his Riviera.

Quick shouted, “You fucking bastards, you fucking bastards!”

This was such an unbearable pain. Quick began to feel that his best friend had been murdered in cold blood. The Riviera was riddled with holes like a cheese grater. Five police officers approached the unmarked vehicle that Quick was placed into. The officer looked like Terry Crews, and Quick paid close attention to him.

"Come on out!" A red-headed officer gave orders to Quick.

This officer stood directly behind the office that looked like Terry Crews. His name was Officer Sanders. Quick noticed his name on the badge that hung around his neck.

Quick stood up once he was out of the car, looked directly into the man's eyes, and said, "How can you be so heartless?"

Officer Sanders said, "I do not give a damn, and you are lucky even to be alive!"

Quick headbutted Officer Sanders, caused a deep cut on his forehead, and even broke his nose. Quick was grabbed by the Terry Crews look-alike and pushed to the ground. Quick was unaware of the two men standing directly behind him. He was still in shock over his grandfather being murdered in cold blood. The two officers finally had Quick under control. Quick noticed a man with a light complexion moving swiftly toward his direction.

"Hello, young man, my name is Tumin. Before I begin, let me tell you a little about my life story."

Quick really did not understand where this high-yellow asshole was coming with this bullshit. Quick had no comprehension whatsoever. As the two men continued talking, he explained that he was from Congo. He continued to explain that his grandfather had been killed in the Congo.

Tumin said, "I came from a place where poverty is high. There's no place for professional-level jobs. So, at an early age, my family fled the Congo, and we moved to Paris, France. As you see, I have a light complexion. I was the lightest African in my village. A white soldier raped my mother. You see, I am trying to explain that I have seen tragedy and experienced it first-hand."

Quick still did not pay attention to this man.

Tumin said, "I see you are not paying attention." Tumin began to laugh.

Quick stared at him as though the man was crazy.

Tumin said, "Well, well, there is that look keeping you alive. It might be my worst mistake, but I will gamble on that. But until your father gives up what he has taken. You both will always be looking over your shoulders!"

Tumin yelled out loud, uncuffing his ass.

People from the community started gathering around the murder scene. People began to cry and scream when they noticed Mr. Gibson slumped over his steering wheel. Quick looked around and noticed Akbar pulling up. Quick felt a sense of comfort when he noticed his dad. Quick began walking toward Akbar.

Before he could walk a few feet away, Tumin stopped him and said, "Oh yeah, I would say sorry for your grandfather, but it would be a lie!"

Tears emerged from Quick's eyes as he took his last look at Mr. Gibson.

"Let's get far from here," Quick heard Akbar's voice say.

Akbar said, "Don't worry, we'll ensure his body is secure."

Akbar grabbed Quick by his shoulders, and they walked away toward the car. Quick noticed Akbar's swollen hand from having punched the glass mirror earlier.

Quick said, "Dad, your hand is severely swollen!"

Akbar replied, “Do not worry about that, son. Let's just get far from here.”

Akbar put the car in drive, not paying attention to the tragedy that had just taken place at his father's store.

Akbar said, “I am terribly sorry Quick, for putting your life in danger!”

Quick replied, “Wow, after all of this, you finally realize what you've done was so selfish! It is too late now! My grandfather is dead! Your father is dead! Even Janice's Aunt Kay has been blown to pieces. Due to your greed!”

Akbar responded, “Quick, you are absolutely right.”

Quick said regretfully, “One of the most influential men in my life has been murdered!”

Akbar understood Quick's life was in danger, including his own, so he immediately called net jet and made reservations to fly out!

“Hello, this is Tiffany. Thank you for calling net jet.”

Akbar said, “Yes. Hello Tiffany, this is Oliver Bryant. I would like to make a reservation for two. My client number is 56- 441-041-7661, with an expiration date of 2030.”

Tiffany responded, “Just a moment, Mr. Bryant.”

Tiffany typed in the information given to her by Akbar.

She said, “OK, Mr. Bryant, how may I assist you?”

Akbar told her, “I would like to fly out to Saint Vincent at 1 A.M.”

Tiffany said, “OK, Mr. Bryant, let's see, here we have a charter flight going out at 2 A.M.”

Akbar said, “2 A.M.? That's OK. That will be fine!”

Tiffany questions, “Would you like a car to pick you up?”

Akbar said, "No, thank you, Tiffany."

Wow! Quick thought to himself. Even though all the bullshit, Quick loved his father dearly. Quick looked over at Akbar, wondering what other secrets Akbar had in store.

Tiffany confirmed, "OK, Mr. Bryant, you have been scheduled for a 2 A.M. flight to Saint Vincent. Please arrive one hour early for check-in."

Akbar thanked Tiffany and told her he needed to make one more phone call. Looking over at Quick, Akbar began making a phone call right in the middle of the conversation.

After a few rings, Laura answered. She said, "Hello, Akbar. What is going on?"

Akbar replied, "I have some shocking news, Laura. My father was murdered today!"

The phone went silent for a brief moment.

Laura replied, "What happened?"

Akbar told her, "Well, the police shot him down in cold blood!"

Laura just went silent, staring at her clock positioned on her desk.

"Laura, Laura!" Akbar yelled into the phone.

Laura said, "Yes, yes. Sorry, Akbar!"

Akbar instructed her, "Laura, I need to ensure my father is properly buried for me. Quick, and I will not be here for that."

Laura said, "Operation Sheepskin?"

Akbar replied, "Yes, Laura, these people will not stop!"

Laura said, "Do not worry, Akbar. I will make sure your father is buried properly."

Akbar hung up without even saying goodbye. He sat in meditative thought. After all, these years, things had seemed so short-lived. The money did not even matter anymore. People have lost their lives over this. But the loss of Mr. Gibson pushed Akbar to the limit. He was going to find these people responsible for this. He promised himself that he would find out. He knew Laura would make sure that Mr. Gibson was buried properly.

Knowing she would answer all necessary questions about why he and Quick would not be present at the funeral. Akbar headed toward Porter Ranch, California, where he had everything stored. Documents, cash, and credit cards. Akbar looked down at his hand.

Akbar said, "I guess my hand will have to wait."

Looking over at Quick, time was of the essence. Things were heating up fast, first thing, the break-in, then Mr. Gibson being killed all in one day. He had to get Quick away, and going to Saint Vincent would be necessary.

Akbar told Quick, "We have a flight going out at 2 A.M. I just have two more stops. I need to dump this car first at a storage facility."

Akbar monitored his side view mirror swerving through traffic, making his way down the 405 toward Sherman Oaks and then to Porter Ranch.

Akbar said, "Kick back, kid. Everything will be just fine."

Akbar pressed his playlist, and the song "*Places and Spaces*" by Donald Byrd blared through the speakers. Before Akbar could get into his music, a call came in from Laura.

Akbar said, "Hello, Laura. Where are you?"

Laura replied, "Well, I am heading to the crime scene."

Akbar responded, "OK, cool, thanks, Laura. Don't call back. I will call you when I arrive in Saint Vincent."

Akbar knew Laura could be trusted if anything had ever happened to him, that Laura would be there for Quick.

Akbar said, “I love you, Laura.” They both disconnected the call at the same time.

Akbar said, “Quick, there is something I need to show you and introduce you to a dear friend. We call this place the end zone.”

After about a 15-mile drive, Akbar exited to the secure location. A place he had established after Laura's boyfriend had been killed. A metal box fence protected the building.

As they pulled into the building, he pulled up to a silver box on the driver's side. Akbar rolled down the window, looked directly into the box, and scanned his entire face. It was a facial recognition security system. The Iron Gate retracted, and Akbar pulled inside.

Quick looked around and immediately noticed a 1972 Harley-Davidson fastback with a custom paint job. Akbar approached a massive building; a great door opened automatically. A tall brown-skinned man with dreadlocks was close by. You could immediately notice tattoos on both of his arms. This man pointed toward an empty parking spot Akbar pulled into.

The spot was next to a silver Audi RS5 with a limo tint. The doors to the Audi had already been opened, waiting for Quick and Akbar to arrive. The engine was already running, Akbar pulled directly into the spot near the Audi, and Quick and Akbar exited the vehicle.

“Shit must have hit the fan!” The brown skin man said.

“Sure did,” Akbar replied.

The man reached into Quick's hand, “Hello, young man, I am Vic. I have heard a lot about you. Your dad and I are longtime friends!”

Wow, you just got more surprises. My dad is in super deep trouble, Quick thought.

Vic turned away from Quick and said, “I know what to do!”

He handed Akbar his duffel bag, and they shook hands.

Akbar said, “Thank you, Vic. See you in the future.”

Akbar smiled and placed the devil bag into the Audi RS5. Quick and Akbar made their way toward the vehicle. Akbar put the Audi RS5 in drive and departed from the storage facility. Vic looked as though his best friend would not return. Quick had been around and noticed that the facility was state of the art. As Akbar drove away, he looked over at Quick.

Akbar told Quick, “Participating in the heist was wrong of me. I just wanted a better life for you. I have one of the best computer science texts in the world in the field of software design. But the dark side has managed to creep back into my life. And now I am paying dearly. Son! Before we take this flight, I want to show you one more place. It is a place that I have set up in Porter Ranch.”

Quick screamed, “Damn it! What else are you keeping from me, pops?”

Akbar replied, “Relax; I will explain everything more deeply.”

Before Akbar could reach the freeway, a call came in for Akbar.

“Hello friend, a voice with a Russian accent came through the speakers in the vehicle.”

Akbar said, “Hello, Oblonsky.”

Oblonsky said, “I see things are getting intense. I received a call from Laura.”

Akbar said, “Yes, very intense, but I will be able to handle it!”

Oblonsky said, “Be safe, my friend, and do not worry, your father will have a nice burial, and I will make sure the business runs smooth.”

Akbar said, “Do not be too hard on my employees, and make sure Seyne stays focused.”

Still monitoring the vehicle mirrors, Quick looked at Akbar and realized he did not know his father. But regardless, He knew Akbar had done his best. After about an hour's drive, Akbar exited toward Porter Ranch at about 90 MPH. Akbar turned onto a straightaway. Quick looked into the distance and noticed a beautiful family home.

Quick thought to himself, *if none of this had happened, would we be here?*

Akbar said, “Well, this is it!”

They pulled up to a single-story ranch-style home where lights were automatically triggered once they drove in. Quick could still not believe that his father had kept this from him. His father had been living a double life. As Akbar pulled into the parking stall and came up to a stop, the Audi RS5 began to descend underground.

Quick thought, *what the F*CK?*

Once they were below ground, Akbar pulled forward and put the Audi in park. As they both exited, Akbar grabbed the duffel bag. At this point, Quick still had not been in touch with Janice. He did not know if he would ever see her again. But for now, calling seemed best. There was so much chaos with everything going on that it was best to lay low. This would help him stay focused and alive, and that is all he kept thinking about, as well as what Mr. Tumin had said earlier.

Akbar shouted, “Quick, come on!”

Quick snapped out of his daydream and started walking toward Akbar. Quick noticed the black door, Akbar put his palm on a small box that scanned his palm print. That triggered the door to open automatically.

As they walked in, a voice came through the speakers that were mounted on the ceiling, “Hello, Mr. Gibson and Abkar.”

Akbar responded, “Hello, Ivy.”

The steel door automatically closed behind them. Quick could not believe it, surprise after surprise. He did not notice Dell servers were mounted against the wall. Flat screens were also mounted on the wall. Quick looked closely, and you could see images of their freeway on the screens—even the streets surrounding the area. Akbar had hacked into the city surveillance system. There were also five giant ceiling fans throughout the home to keep the home cool due to the number of computers.

Akbar noticed how Quick observed everything.

"Those fans are top of the line. They have sensors that automatically control their speed without me having to adjust them. How about we get something to eat? There is so much I have to show you and tell you, son."

Akbar and Quick walked toward a solid oak table. There were two chairs at each end of the table.

Akbar said, "Have a seat, son."

Quick pulled out one of the chairs from the oak table for himself.

"First, let me start with this, as technology has progressed, we have also progressed. Quick, a longtime friend and I built two artificial intelligent robot humanoids. What is so different is that these two can feel emotions, something scientists have not been able to accomplish yet."

Quick said, "So, you are telling me you invented a robot, human?"

Akbar said, "Yes, that is what I have done, and society was not ready for this technology."

Quick commented, "I never thought in a million years that my dad could or would build an intelligent robot or a humanoid. Whatever you want to call it."

Akbar replied, "Well, Quick, I did! The heist I was telling you about. As you can see, they are killing off our family. Those shipments consisted of military computer chips developed just for that purpose, and those individuals wanted them back at any expense. There was an official high officer, Lieutenant Lawrence Johnson. He was head of the operations of the Artificial Intelligence division of the United States, Rogue Army. A very deep and underground unit. They were known for trading with Russia and China, and who knows who else. As you have noticed, I quickly transformed this entire home into an intelligence fortress. I hacked into surveillance systems in our area."

Quick expressed, "Damn, dad, I never knew, but look at it, I mean, what did this cost you?"

Akbar walked over to the fridge and grabbed himself and Quick a bottle of Evian water as Akbar sipped on his water bottle.

He looked over at Quick and said, "There is something I would like to show you, son. Follow me."

As they walked away from the table, Quick followed behind. They approached a full-length mirror. Akbar touched the middle of the mirror, and it opened the wall behind the mirror. As the two entered the small space, a spiral staircase went down approximately twenty feet to a lower level.

Once they reached the bottom, there were two artificial intelligence humanoids. One of them was a male, and the other was a female. Quick's eyes popped out of his head practically. The boy was in shock by what he was observing. He could not believe what was in front of him.

Akbar said, "Quick, you do not have to be afraid. These two will protect you, and it is time for them to be activated. Come over here."

Akbar pointed in a direction away from both of them. They arrived near a family portrait, and behind this portrait was a small safe mounted on the wall. The picture was of Mr. Gibson with the family. A tear came down from Quick's right eye. Quick looked at the picture, and Akbar pressed the code 619915. Inside there was a flash drive. Akbar grabbed the Flash drive.

"Quick, on this flash drive. You will learn how to operate these androids. The only person who knows about this location is Laura and Vic." Akbar closed the safe.

Quick said, "Dad, I want to know more about this Lieutenant Lawrence."

Akbar replied, "Well, Quick, Lieutenant Lawrence is behind all of this as well as a man named Santiago. That shipment belonged to them, but there is more to it. Even China has a hand in this. Could you imagine if they created artificial intelligent terrorists? Hell, no, not on my watch! These machines should be used for the greater good of humanity. But I will explain it to you later; it's time we sit down and have something to eat, son."

Chapter 4: Coming Home

"That's right, girl. Do your *thang*!" Quick yelled out from the stadium stands.

It had been two years since Quick and Janice had seen each other. Akbar and Quick had spent two years in St. Vincent, and they had pretty much dropped off the face of the Earth. It had also been two years since Mr. Gibson was murdered in cold blood.

Quick's entire attention was directed to the people responsible for killing his grandfather. He thought about Akbar's definition of "attention" and realized what he would say, Attention is the ability to focus on what we perceive. We can perceive a lot of things simultaneously, but what stays in the forefront of our mind and knowing the reality of what's happening is attention.

"Baby, baby!" Janice had been calling for Quick.

Quick immediately snapped out of his daydream. He looked back at Janice, smiling; he kissed her on the cheek.

"Damn, baby. You did really good in that one hundred meters," Quick appreciated Janice.

"Thank you. We have a lot of catching up to do." Janice replied as the two exited the stadium bleachers.

A tall, burly man followed them. But he did not notice that Gus, one of the artificial intelligence humanoids that Akbar had built, kept security for Quick.

"I will race you to the car," Quick yelled out.

"I am good. My legs are killing me."

As the two continued walking toward the vehicle, Quick pushed a small button to unlock his vehicle.

"Do not look back," Quick instructed Janice.

"Just relax and do as I say, Janice. Okay?"

"Okay."

Gus grabbed the tall, burly man around his neck and started choking him until he was unconscious. Janice didn't even realize what was going on. Gus looked around and saw that no one had noticed what had happened.

A silver Mercedes Benz cargo van pulled up alongside Gus. As he threw the man inside the cargo van, the door automatically closed behind him. Gus immediately zip-tied his wrist. He then texted Quick, 'mission complete.'

"Baby, I will have to end our date short. Something important came up." Quick said.

"Why?" Janice inquired.

"Well, there is something my dad needs me to do," Quick explained.

Janice knew Quick had changed in those two years. She looked at him and noticed a grown man as she smiled. Quick turned up the music. The song, '*Don't fuck with Shady'* by Eminem, blared through the car speakers. The two bobbed their heads at the streaming music in the background. Janice wanted to be involved in his life more than she currently was.

***

Gus slapped the man tremendously hard. He immediately awoke after being slapped.

"Who do you work for?"

"Fuck you! Fuck you!" The man replied, spitting in Gus' face.

Gus grabbed him around his neck, spread his eyelids, and placed black contacts over his pupils. Everything went dark, and he began to scream. Gus punched him, knocking him completely out.

***

After a 20-minute drive, Quick pulled up to Janice's house. Looking over at Janice, he noticed a tear coming down her cheek.

“Don’t cry, baby. I will call you as soon as I'm done,” Quick consoled her.

“All right, but promise me you'll be careful.”

“I promise,” Quick replied.

Quick gave Janice a passionate kiss before she exited his vehicle. Looking back, she noticed a fire in Quick’s eyes. He then quickly pulled away from the curb. He could not wait to reach the end zone since it would be his first time torturing a person. Akbar had taught Quick so many things over the years.

Gus pulled into the end zone, and the steel gate retracted automatically. You could hear the motor as the gate closed behind them. A scream came from the back of the van.

“Okay, okay,” he said. “I will tell you whatever you want to know.”

He understood wherever he was; it was his last days.

“Well, sir, it really does not matter. I already know you work for (S.O.C.O), a special forces security. Security force special operation. Too bad your last day will indeed be here.”

Pulling into the garage, Vic walked toward the cargo van. The van's side door opened. He immediately grabbed the man wrapping a cable around his neck and snatching him so violently that the wire created an inch-deep cut. Gus and Sparta exited the cargo van. Sparta is an artificial humanoid as well. Vic placed their victim on a steel slab like a gurney you would see in a morgue, laying him face up.

“Excuse me, please don’t do this, man,” the man cried.

Vic remained silent, not paying attention to what he was saying.

Gus and Sparta looked over, seeing Quick exiting his vehicle and walking toward them. All Quick could think about was how he would end this man's life. As he walked in, two glass doors opened. Quick could see Gus and Sparta standing next to Vic. Before entering, he looked into a facial recognition scanner screen, and the door opened. Vic turned and noticed Quick had extended his hand.

Vic extended his hand to shake it and said, “Hello, young man.” He smiled, “Well, we found who sent him. S.O.C.O, a special forces task force. They deal in high-tech weaponry. It seems as though he will talk.”

As soon as he pulled through the gates, Gus mentioned, “He pleaded for his life. I know he will talk.”

“The bastard better talk,” Quick replied. *This will be for my grandfather, aunt Kay, and my father*, he thought. He took this first kill personally.

“Let's get busy,” Vic explained.

Inside their interrogation room, numerous monitors were mounted around the room. The shop had been state-of-the-art, but the aquarium with an albino king cobra stood out in the corner.

“Okay, tell me what you want to hear,” Quick explained.

“My name is Mastroni. I was sent to do surveillance, and there will be many more after me.”

“Well, I will worry about that later, but I promise you the people that sent you will be dealt with even worse,” Quick explained.

He then pulled a 0.45 caliber gun out of his waistband. Vic had grabbed a burlap sack and placed it over his head. Quick put his 0.45 caliber against the man's head and pulled the trigger.

Boom!

Blood soaked the burlap sack. Quick looked directly into Vic's eyes.

"I will be back," Quick said, putting his 0.45 caliber back in his waistband.

Vic walked toward the wheelbarrow outside their interrogation room and returned with Mastroni's body. Gus and Sparta followed Quick, walking toward a Mercedes-Benz wagon. As Quick sat inside, leaning back, thinking about his dad, *how would Akbar feel about Quick killing Mastroni?*

Their stay in Saint Vincent had been cut short because Akbar had died in a zip line accident. The harness had snapped, and abkar plunged to his death. Quick never had a chance to view Akbar's body since Saint Vincent Police Department refused to let him see it.

After a week in Saint Vincent, they issued him a death certificate. Quick retained everything his father had taught him. Akbar taught Quick economics, history, technology, and self-defense; he wanted to ensure Quick was prepared for society. And as per his expectations, Quick excelled in everything Akbar demanded of him.

As the three drove down the 405 freeway, Quick looked at the car passing him, missing the two most influential men in his life.

Quick dialed Janice from a secure line, and after a few rings, Janice answered, "Hello, baby."

Quick replied, "Sorry for not being able to stay longer with you." Janice had no idea what was going on. She loved him a lot. She did not even know that Akbar was no longer among them.

"Do not worry about it, Quick. I love you." Janice said.

“I will make it up to you, I promise,” Quick replied.

“Sure.”

They both ended the call.

After driving for hours, Gus pulled into the ranch home. Quick looked out into the vast amount of land that surrounded their home—remembering the first time Akbar surprised him, pulling up to the secure location years ago. The three exited the Mercedes-Benz wagon, and Quick walked toward a small box alongside a steel door; placing his eyes against the box, it scanned Quick’s face.

As the door automatically opened, walking in, he noticed how organized the ranch home was. Quick walked toward the surveillance cameras and pulled a small keyboard toward him. With a few keystrokes, he pulled up surveillance footage for the past six hours. Observing the video, Quick noticed an unmarked car at the rear by the minimart. Suddenly, a flashback hit him. He noticed the face of the person exiting the vehicle.

Alvarez, Quick thought, *how and why would he be so close to their secure location?*

He immediately hacked into the cameras of the entire surrounding area, monitoring Alvarez’s unmarked vehicle. He noticed his unmarked vehicle heading in the direction of the ranch. Alvarez never turned down their service road.

“What the fuck,” Quick yelled out. “How the hell could he be so close.”

How!! How!! Quick kept replaying in his mind. He moved away from the surveillance cameras, and his blood started boiling seeing Alvarez so close to his home. Alvarez had been one of several who fired their weapon, killing Mr. Gibson.

“Gus,” Quick yelled, “We need to find out how Alvarez came so close.”

Gus walked toward the surveillance cameras, zoomed in, took a facial recognition of Alvarez, and programmed Alvarez into the database. Alvarez's home address automatically stored in Gus's memory. Gus stepped back; you could see the rage that burned through Quick's face. Knowing whoever sent Alvarez this close had surveillance on Quick.

But how? Quick had just returned from Saint Vincent.

Gus headed to pay Alvarez a visit.

***

Major Conroy always kept his 45-caliber close at hand. He always felt someone would end his life very violently. Conroy could never relax. Even when it came to relationships, he would call escorts only. He would strip down naked, waiting for his sexual encounter, not knowing Sparta would be his last.

As Sparta made her way through the hotel lobby, men instantly turned their heads. She was tall, slim, beautiful, and had a dark complexion. Her smile illuminated the entire place as she gracefully made her way through the hotel lobby.

Smack!

The sound of a woman slapping her husband as he stared at Sparta walking through the lobby. Sparta pulled out a small key card making her way toward the hotel stairway. As she swiped her keycard, the stairwell door opened; she made her way toward the fourth floor. Before exiting the elevator, Sparta adjusted her name badge.

This is where a pinhead camera had been installed. Gus could see everything from his laptop computer as he observed from a nearby V70 Volvo. He hacked into the hotel surveillance cameras; he had installed a software where the cameras would depict Sparta as a 60-year-old man. Sparta exited the stairwell onto the fourth-floor hallway. Sparta made her way toward major Conroy's hotel room softly.

“Come in,” Conroy responded.

Pulling the handle, Sparta entered a pitch-dark room. Her night vision automatically engaged. She could see a major convoy pleasuring himself. Major Conroy made it a habit of making sure when women would come in and see him, all the lights would be off, making the women follow his voice. Sparta stood still, waiting for Conroy to give directions as she undressed, and the only thing she kept on was her combat boots.

“Come here,” Conroy spoke.

Walking toward Conroy once in striking distance, Sparta kicked Conroy in the face. Numerous teeth fell out of his mouth; blood splattered everywhere; Conroy's body fell limp. Sparta grabbed the remote that was sitting on a nearby dresser. The light turned on by pressing a green button on the remote. Looking down at Conroy, still unconscious from the blow Sparta gave him, she picked up her black bag, where she had stored a few items.

She then grabbed a stud finder, nails, and a small sledgehammer and walked toward the near wall, taking her stud Finder to find the location of the studs.

Beep, beep.

She ran her stud Finder along the wall. Once she found the studs, she punched a hole in the wall. She walked over to Conroy, slapping him across the face, awakening him. He looked up, shaking his head as though someone had hit him in the head with a hammer. Conroy could taste blood inside his mouth.

“What, what?” Exclaimed Conroy.

“Well, first and foremost, some years ago, a young mother had been killed in a car bombing in long beach. A gentleman by the name of Alvarez gave you up,” Sparta explained.

“I don't know anybody by the name of Alvarez,” Conroy yelled.

Sparta grabbed Conroy by his ankles, dragging him over toward the hole in the wall. She grabbed his throat and gave him directions to position his hands. Once set, Sparta grabbed her small sledgehammer and nail, then hammered one nail into his hand, going through directly into the stud.

Major Conroy let out an extremely loud scream. She punched him in the face, knocking him out. She held his other hand, slamming another nail completely through, and walked back toward her black bag, reaching for her ammonia pack. She looked around the room and noticed Conroy's laptop. Pulling out his flash drive, Sparta walked back toward him, putting the ammonia pack directly under Conroy's nose.

He inhaled the ammonia pack, and she continued, "Now, back to my question. Well, your employer Akbar had stolen some very valuable merchandise."

"Computer chips. They paid me to make his life miserable," Conroy explained.

Sparta knelt, pulling Conroy by his ankles and ripping him from the wall. Conroy passed out from extreme pain. She dragged him toward the hotel bathroom as he lay unconscious beside the sauna. She turned the knobs for water to fill the tub until it was full. Conroy awoke, and Sparta turned around and grabbed him, abruptly throwing his body in scalding hot water.

"Ahhh…" Conroy screamed, but his scream was muffled when Sparta dunked his entire head underwater, splashing water everywhere.

Conroy breathed for his life but lost the battle as his body lay lifeless in the tub. Sparta dropped two acid tablets in the water as they dissolved, and within minutes, it began to break down Conroy's flesh.

She then exited as the room filled with an unbearable smell. Grabbing her clothing and tools and not to forget Conroy's laptop while exiting the hotel room, Sparta still had a large amount of blood on her. A young woman noticed blood on Sparta when she made her way through the lobby. It shocked the young woman.

Sparta walked toward Gus, who was waiting for her downstairs. He noticed blood on Sparta. Even though Gus had been a humanoid, he was highly intelligent and knew that was not good. Sparta's completed her first mission but made a huge mistake. Gus Drove away toward Interstate 5 when a call came through.

“Hello, Gus. How did everything go?” Quick asked.

“Things went just fine, but a mistake….”

“Mistake? What kind of mistake?” Quick inquired.

“Well, Sparta did not clean herself up before exiting the room,” Gus explained.

“Okay, okay, just get back here safe.” Quick hung up.

That is what Quick could not afford, Sparta getting caught. Quick would have to shut her down for a little, just in case things got heated up.

Quick knew a journalist by the name of Bernice Jamison, who would know if any witnesses gave a description of Sparta.

After a few rings, Bernice picked up, “Hello!”

“Hi, Bernice. This is Quick.”

“What is up, young man?”

“Well, Bernice, I have a little situation. If you hear anything, please contact me,” Quick explained.

“Man down?” Bernice replied.

“Yes, man down. If you hear anything, let me know.”

“Okay, I will.”

Quick understood Conroy's foot soldiers would be asking a bunch of questions. Whoever got that close to Conroy had done their homework.

***

You could hear glass clanging in the hotel lounge as a tall man yelled, "Another round."

"No, no." A slender woman responded. "It is time to end this," she ordered.

The small military team exited the lounge, but the slender woman left a huge tip before leaving and walking toward Conroy's room. Before the team could reach Conroy's room, a call came in from the fourth floor.

"Hello, Ms. Lewis. You told me to call if anything strange happened or if I noticed anything. I was pushing my cart and noticed blood on major Conroy's door." Sarah explained.

Ms. Lewis hung up, got her team's attention, and pointed toward the hotel elevators. The security team rushed toward the elevators, pushing hotel customers out of their way.

"Watch out," one of the team members yelled out, damn near pushing a young couple into the wall.

"What the fuck!" A young man yelled out.

The team paid no attention as Ms. Lewis's team passed the fourth floor, but she still had no response. Ms. Lewis noticed a light coming from the bathroom. Moving closer, Ms. Lewis noticed blood on the bathroom floor. Luis entered and noticed a hot tub full of blood; the smell was unbearable.

One of her team members vomited throughout the hotel bathroom. Ms. Lewis told one of her team members to secure the hotel room.

"Hardgrove!" Ms. Lewis gave directions. Hardgrove was a short, muscular man with dreadlocks.

"Make sure no one exits that elevator."

Lewis gave directions as he grabbed her phone, calling her boss, Santiago. After a few rings, Santiago answered.

"Hello," Santiago replied.

"Someone has caused harm to your cousin," Lewis explained.

"What do you mean?"

"Well, sir, he is nowhere to be found, and there is a hot tub full of blood."

"Okay, clean up the place and report to me immediately," Santiago explained.

"Taylor Jefferson, make sure this room is spot clean," Lewis ordered.

Taylor made a phone call and sounded as though he had put in a fast-food order to go. Lewis pointed toward her team as they exited the hotel suite.

Lewis thought to herself, *who would have wanted to kill Major Conroy? And was that blood Conroy?*

As she was walking out, Lewis's phone vibrated. It was her loyal friend Smith.

"Hello, Smith. I need your assistance cleaning up."

"I am on my way," Smith replied.

Entering the elevator, Lewis could not believe someone had been killed on her watch. She had been doing security for 35 years. Lewis and her team exited the hotel.

Pulling her Versace glasses out of her jacket, she walked into the sun and toward a fleet of black-on-black Range Rovers. Lewis was wearing a tailored made suit with high-end custom tennis shoes. She loved tailor-made suits and never wore high heels.

Sitting in the rear of her Range Rover, it still did not sit well with her that major Conroy had gone missing on her watch. Santiago loved Conroy. She thought Lewis was one of the best private security companies in the world. With that thought, a smile appeared on her face.

Conroy got what he deserved, especially the way he treated women. But she was hired to protect, not to get emotional. One of her team members closed her door. She thought about major Conroy making several advances toward her, but that had been short-lived. Louis damn near broke major Conroy's finger, but reality kicked in; she might not work in security again.

"Let's go, king," Lewis gave orders, and the Range Rover headed toward Malibu toward Santiago's home.

She dialed Smith once again. "Do not forget those security surveillance recordings," Lewis reminded Smith.

Smith and Lewis were very close friends, and they had become very wealthy, but they still loved their first job protecting clients. Lewis knew that she could count on Smith for anything. After a 30-minute drive, Smith pulled his four-by-four Jeep Cherokee into the hotel's underground parking lot.

Finding a nearby parking spot, Smith walked toward the service elevator. He heard the ding without pushing the elevator button. Smith walked directly into the service elevators, pressing a button that reads security.

After a few seconds, Smith reached the floor, where it read security office with an arrow pointing to the left. The aroma of onions filled the hotel hallway; Smith noticed a small camera once he exited. He continued in the direction of the hotel security office. Approaching a metal door, Smith knocked.

"Hello, may I help you?" One security officer responded.

From behind the door, Smith talked into a small speaker outside the security office.

“Who is it?” Byron replied.

Byron observed security cameras while Smith flashed his badge.

“Hi, I am detective Lawrence from the Los Angeles Police Department.”

“Yeah, what's up?”

“Well, I need to see your surveillance footage regarding identity theft,” Smith explained.

Byron, who was new on the job, never noticed that Smith’s badge didn’t have LAPD markings. Byron looked over at Morgan, “Hey, this guy wants to see surveillance footage.”

“Who is he?” Morgan asked.

“Well, he is an LAPD detective.”

“Okay!” Mark replied.

Morgan continued talking on his cell phone as Byron pressed his button, unlocking the door and letting Smith enter. Smith looked around and noticed a very well-organized office. Smith closed the door behind him as Byron reached out to shake Smith's hand and greet him. Smith never acknowledged Byron's handshake.

Smith made his way toward the hotel security monitors, “Sir, Sir, excuse me,” Byron was trying to get Smith's attention.

“I do not have time. I need all surveillance footage,” Smith looked at the logging of his heritage military watch. “I need all footage from six hours ago,” Smith explained.

“Sure,” Byron replied.

Byron knew Smith's visit was very serious just by his facial expression. Smith's phone began to vibrate; Lewis had been calling, Smith answered.

"Have you found anything yet??

"No, I am working on it now," Smith explained.

"Sure, I am on my way to Santiago now," Lewis replied.

Lewis ended the call. Smith already knew what to do after he found the footage. Smith looked at both security guards.

"Okay, gentlemen, let's get to work."

Smith took over the security monitors and zoomed in on the hotel lobby observing all guests. Smith noticed a tall attractive woman who stood out amongst the rest, nowhere on any camera, just an older gentleman in his late 60s. After looking over all the surveillance cameras, they still couldn't see Sparta.

"There," Smith said as they noticed a gentleman walking directly into major Conroy's hotel room.

They monitored Conroy's door. After about 20 minutes, Major Conroy's door opened.

"Do not take your eyes off him," Smith yelled out.

The three closely watched the surveillance cameras.

"There, there..." Byron screamed as they noticed the lobby exit door.

Sparta appeared once again. Sparta walked through the lobby, exiting the hotel.

"Do you have any cameras outside?"

"Yes," Byron replied.

They notice Sparta walking toward a silver V70 Volvo.

"Get that license plate number," Smith replied.

“Yes, sir,” Byron replied.

Smith suddenly grabbed Byron around his neck and snapped Byron's neck with one twist. Morgan looked over and was shocked as he watched Smith snap Byron's neck.

“Please don't do this,” Morgan pleaded with his hands in the air.

“Stay silent,” Smith said as he did the same thing to Morgan, snapping his neck as well; both security guards lay dead.

Smith reached into their pockets, removing their wallets. But before exiting, he removed the flash drives and hard drives. Smith called Lewis.

After a few rings, Lewis answered. “Well, what do we have?”

“I have a lead!” Smith replied, “Make sure the fourth floor is clear completely.” Smith hung up.

Smith exited, looking both ways, making sure everything was clear.

***

“This sun is burning my ass off,” Laura exclaimed.

“Don't worry, I have some handy,” Janice smiled and started looking for something in her bag, then she took out a bottle of sunscreen and handed it to Laura.

As the two women continued their conversation, Janice pulled her glasses down when she noticed Quick walking in her direction.

Laura looked at Janice and said, “Girl, I wish Quick would have continued his education.”

“I know, hopefully, one day he will continue,” Janice replied.

They both lay back and started soaking in the sunrays while listening to music. They also kept an eye on Quick, who was moving through the crowd. Quick had such a radiant smile, and his pearly white teeth made it even more perfect.

"How is my Angel smiling," Laura giggled.

Janice immediately jumped in, grabbing Quick around his waist and giving him a long passionate kiss.

"OK. OK. Go somewhere else with that," Laura laughed.

Quick and Janice continued kissing, but a few seconds later, Quick pulled back.

"Can I have a word with you?" He asked.

Janice turned with her beautiful smile and said, "Laura, we will be back in a few minutes."

Laura nodded as Quick and Janice walked toward the back of the yacht.

"There is a small boat waiting; let's go out of here," Quick said while pointing toward the boat.

They both started walking, and Quick grabbed Janice's hand to escort her toward the small awaiting boat.

One crew member unlocked the small boat for them, and Quick started the engine.

"Quick, what about our guest?" Janice asked.

"Do not worry about them," he laughed.

"It's just you and me," Quick continued.

After about ten minutes, Quick made his way to a small, secluded island. He had been planning this surprise for a while, and when Janice learned about it, she was amazed. She could not believe that Quick had organized a party for her. However, suddenly without notice, Quick stopped the engine.

"What's wrong?" Janice inquired.

"Nothing, my love, but from here, we must swim to that small piece of land," he replied.

“What? really?” Janice laughed.

“Yes, really!” Quick giggled and jumped into the freezing cold water.

“Come on, I will race you,” Quick yelled while swimming his way through the waves.

Janice jumped in, and they both raced to shore. After a brief swim, they both reached their destination, panting. Quick reached out for Janice's hands, assisting her out of the freezing water. Janice smiled while wiping water off her face.

After a few minutes, she noticed a gazebo and a table set in two very eloquent places sets. She held Quick's hand and walked with him to their waiting table. As they walked, their feet sank into the sand.

Quick pulled back a chair for Janice and directed her to sit down.

“Thank you,” Janice said, taking her seat and smiling with a tear falling down her cheek.

Quick wiped off her tear with his thumb and kissed her forehead.

“Don't cry, love. It's your birthday,” Quick said.

“There are a few things that I need to explain. It is important, and I promise it's not going to rain on your head,” he continued.

He took his seat at the other end of the table, paying close attention to Janice.

“Well, first thing, I found the person responsible for killing your Aunt Kay, and he is no longer alive,” Quick revealed.

Janice's eyes and face looked shocked at what she had just heard.

“I knew all along, Quick. Just the way you had been acting. I have always wanted to be a part of your secret life,” Janice explained.

“But still, thank you,” she smiled.

“You're welcome. I always knew one day you would find out but not that fast, I wish your cousin J.C. could hear this, but one day he will,” Quick said.

Janice began to cry, and Quick leaned forward to comfort her.

“Baby, me and my father owed that to you and the family. My father was part of a heist, and the merchandise belonged to some dangerous people. Innocent people, including my grandfather and your aunt, have been hurt and killed because of this heist.” Quick explained.

“So, it is over, or do we still have to watch our back? Janice asked.

“They are very powerful people, so I am still alert, but we have the advantage,” Quick replied.

“What type of advantage?” Janice questioned.

“Well, the shipment carried computer chips for a special military operative unit. They specialize in making artificial intelligence humanoid robots and information warfare,” Quick explained.

“What do you mean information warfare?” Janice asked.

“Well, it aims to influence decisions, hacking computers, drone warfare, and artificial intelligence. These robots are so human-like, and they are or could be used to influence the governmental positions,” he expounded.

“My father never knew those particular chips were so important. So, after your aunt Kay had been killed, and with us being harassed, my dad built two robots.” Quick continued.

“Wow!” Janice replied.

“Hold on a second,” Quick grabbed his phone and dialed 6177. Two minutes later, a tall, brown-skinned, muscular man walked toward Janice.

“Hello!” He said, standing next to Janice.

Then he put his hand in his pocket and pulled out a small box. Janice looked as he placed the small box on the table.

"Go ahead, open it," Quick asked while looking at Janice.

Janice grabbed the box and opened it. She could not believe her eyes and was in complete shock, pleasant shock. "Will you marry me?" was engraved inside the box, and Janice was constantly looking at it in amazement. Janice couldn't believe that Quick would ask for her hand in marriage.

"Yes, yes, and yes," Janice screamed.

Quick looked over at Gus and smiled. "Well, that is one of the robots my dad had built," Quick said.

Janice looked over and replied, "Wow, so human-like. I would have never known.

"Baby, he will protect you. Also, Sparta, the second robot, is female," Quick explained.

Janice nodded her head and said, "Ok."

"After grandpa's death, my dad and I disappeared. He taught me so much. This journey has been a struggle. Here I sit, not older than 19 years, having the responsibility that would have broken down most teenagers. I just want you to know I have always loved you and want you for every part of my life." Quick said.

Then he lifted his wine glass and wished Janice Happy Birthday again.

Janice could not stop smiling. She loved Quick with all her heart. As the sun began to set, they both danced to the music, feeling great just being in the company of one another.

Quick's phone rang as they both enjoyed the best moments of their life together.

He looked at his ringing phone, turned to Janice, and said, “I just want you to know you are my favorite girl.”

Quick put the incoming call on hold and played the voice message that Laura had just dropped. “Hello lovebirds, your guests are waiting, “Laura's voice came out of Quick's phone speaker.

“Who was it?” Janice asked.

“Well, Laura said the guests are waiting,” he replied.

“Gus, please clear up. We have to get back,” Quick instructed.

Quick grabbed Janice by her hand, and they both ran back toward the ocean water, laughing all the way. Loud splashes could be heard as they swam back toward the small boat.

“Last one to the boat is a rotten egg,” Quick yelled out.

Racing toward the awaiting boat, Janice made it first as she climbed in, and the rotten egg, Quick, climbed in after giving Janice a kiss.

“I love you, baby,” Quick expressed.

Then he started the small engine and headed back toward their yacht. Heading back, Quick stared at Janice with love in his eyes, as she would soon be his wife. After a few miles, they finally arrived where their Yacht was waiting.

As the two crew members grabbed ahold of the front end of the small boat, one crew member reached for Janice's hand, helping her aboard.

“Watch your step,” Quick said.

As Quick and Janice made their way aboard, Quick pulled Janice close, gave her a long passionate kiss, and whispered, “I love you.”

Janice blushed.

After getting off the yacht and walking toward their guest, they heard a guest yell, “Happy Birthday.”

Janice grabbed her chest with her left hand; the excitement made her exhale. Champagne glasses could be heard popping as the DJ suddenly stopped the music to make a special announcement.

He called Janice's name and screamed, "It is a special announcement to the birthday girl as we have a special surprise."

Janice looked around and noticed a crowd of people was opening a pathway. There stood her father standing six foot four inches tall and smiling at her. He was muscular and had dark skin tone, with salt and paper hair. Janice could not believe that her father was there as she had not seen her dad since being a baby.

Wow!

Janice thought to herself, and tears of joy came down her face. She never thought she would see him again. *But why didn't he tell me he was coming,* she thought to herself, but whatever the reason, she would not have to see him in that confirmed place again.

There he stood, dressed in a Versace suit, looking extremely handsome. Janice walked toward her father, opening her arms and giving him a warm hug.

"How long have you been home?" Janice asked.

"I've been home for 45 days now, I just wanted to surprise you, and it worked out perfectly," he answered.

"What!? you been home for 45 days?" Janice said angrily while looking at her dad.

"Calm down, please. Let me explain," Jasper grabbed Janice by her hand. "Can we take a walk?" He requested.

As Janice nodded, they both made their way through the crowd. *You do not want to fuck with* by Eminem was playing in the background as they both walked. Some wished Janice happy birthday as she made her way toward the deck with her father.

Quick noticed Jasper and Janice exiting the deck, so he stood at a distance, thinking about how he would explain to Janice that he knew her father had been released from prison. He distracted himself from the thought by thinking he would deal with it later.

Janice and Jasper made their way toward the guest lounge, and Jasper opened the lounge door for Janice, making her way toward a comfortable leather couch. Taking her place, Janice fixed her hair.

"Well, how do I start?" Jasper wondered out loud.

"I could not tell you. I am tired of the surprises, but I guess they are needed." Janice replied.

"Yes, Janice, they are needed. What do you want to know about technology and genetic engineering?" He asked.

"Not much, just a little question. Well, some Chinese researchers recently edited the genes of a non-viable human embryo. What they did no human has done, mixing genetic engineering with robots, making them as human as possible?" Janice inquired.

"Akbar found out and came up with their blueprint, the first humanoid to feel emotion," Jasper pulled the bracelet from his pocket, and Janice looked.

"What's that?" She asked while pointing toward the bracelet in her father's hand.

"Well, dear, I'm sure Quick made you aware of the two robots," Jasper said.

Janice replied, "Yes, dad, he told me all about them."

Well, this is a tracker to keep you safe. There is a sensor inside for you and one of the robots to assist you in time of need," Jasper explained.

Janice looked surprised, "Thank you, I really appreciate this, dad," she said.

Jasper attached the bracelet to Janice's wrist while Janice constantly looked at his face.

After a few seconds, she said, “Dad, I truly forgive you.”

Jasper embraced a wide smile on his face and hugged his daughter.

As they both made their way back to the party, the song *lemonade* by Beyoncé played in the background. Janice noticed Quick looking out into the distance; the full moon had glistened off the ocean.

“Hey, young man,” Jasper said as he touched Quick's back, tapping him.

Quick had been caught off-guard.

“Do not worry, young man. It's just me,” Jasper ensured.

“Quick, there is something I want to tell you. Your dad raised you very well, and I am proud you are with my daughter,” he continued.

“Thank you, Jasper, I appreciate that,” Quick replied.

As the party continued, the noise and laughter throughout the yacht could be heard. Quick's phone rang with the ringtone *say it to my face* by Drake. Quick noticed that Vic had been calling him.

“What's up, Vic?” Quick asked.

“Well, I need to meet up with you in Virginia,” Vic answered.

“OK, I will call you once we dock,” Quick said and hung up the call.

“Sorry for that interruption, Jasper,” he apologized.

“No problem. Is everything OK?” Jasper questioned.

“Yes, but I will have to call it a night,” Quick said, then started walking toward his sleeping quarter.

He reached for his phone and dialed net jet.

After a few rings, a deep voice answered, "Thank you for calling net jet. This is Jermaine Roland. How may I help you?"

"Hello, Jermaine, my name is Robert Coleman," Quick said.

"I would like to make a reservation," he continued.

"Sure, Mr. Coleman. What is your net jet account number?" The voice from the other side asked.

"It is 466 - 15436114, expiration date 2037 April," he replied.

"Thank you, sir. Please hold one minute," Jermaine said as she entered all the necessary information.

"OK, Mr. Coleman, where will you be traveling to?" She asked.

"From Florida to New York City. I would like a flight for 8 P.M.," Quick said.

"Hold on a second," she said and looked up all available flights.

"Mr. Coleman, we have a flight leaving at 8 P.M., arriving in NYC at midnight. Also, Mr. Coleman, will you need a car to pick you up?" Jermaine asked.

Quick thought for a few seconds and then said yes.

"OK, what is your location?" She said.

"Tampa, FL pier 6060, on a yacht called sundress 19," Quick explained.

"OK, Mr. Coleman, you are all set. Thank you for choosing net jet," Jermain concluded.

"No problem," Quick said as he hung up the call.

He lay back on the bed kicking his sandals off, but before he could undress, Janice barged in and jumped directly on his lap. Quick laughed as they both wrestled, made love, and finally fell asleep.

After only a few hours of sleep, Quick turned toward Janice, still sleeping, and gently kissed her forehead. Janice made a soft sexy moan.

"Do not get up, baby!" Quick said, holding Janice tight.

He rolled out of bed to prepare for his flight. He began packing a medium-sized duffel bag, Tommy Hilfiger pants, a leather Nautica belt, Belstaff Roadmaster jacket and putting the small bundle of clothing in his bag. Before leaving the room, he looked back at Janice. Then he made his way down the narrow hallway.

Laura suddenly surprised him in the hallway. She stood before him and said, "Yeah, I already know we have been heading back toward Tampa for the past four hours." Quick smiled.

"That's why my dad and I love you so much," Laura signaled him to come with her.

They made their way toward a small boardroom with a large oak wood table and eight luxury chairs. Quick pulled the chair back for Laura, and Laura appreciated his gesture.

"Thank you, Quick, you are such a gentleman," she said.

Quick took a seat right next to her and started talking, "Well, Vic called me, you might be aware of that already. He wants me in Virginia ASAP."

"Yes, I know I am fully aware of what is happening. A federal agent has been compromised, and Vic is getting information from him. About operation' sheepskin," Laura blamed.

"Ok, so why do I need to be there?" He asked.

"Well, Vic wants you to hear personally. Sit down and hear everything first-hand about operation sheepskin. We only have bits and pieces. Quick you have been introduced to some extraordinary things. Being a young black male with so much knowledge in math and computer science, your father made sure you would be prepared for this generation of technology," Laura explained.

“The advantage is yours, even in the darkest days, just remember how Akbar and your Grandfather raised you,” she continued till a knock on the door interrupted their conversation.

“Come in,” Quick allowed.

Janice walked in, wearing a pair of sweatpants and a long black T-shirt made by the city breeze custom T-shirt company. The shirt read 360 degrees.

“Have a seat,” Quick said to Janice.

Laura continued to speak as she referred to movies and old cartoons such as the Terminator and Jetsons, speaking about writers' imaginations that turn society into reality, which had a grim society for the unimaginable.

“Even speaking on how some in society became fearful of artificial intelligence, especially robot humanoids thinking one day they would take over the world,” Laura said.

“They were made for battle. My father always believed that science and technology are engines that drive civilization forward,” Quick replied.

“Quick, your father would be so pleased with you; he has taught you incredible things,” Laura said.

“Thank you, Laura. You and Vic have been great mentors in my life, and my father would be well pleased with you both as well,” Quick showed his gratitude.

Laura looked into Janice's eyes and then also looked at Quick.

“Jasper really admires you Quickly. I am sorry about your short-lived life as a youth, especially not being able to finish school and play the sports you genuinely enjoy,” Laura said.

“Never apologize for anything. I have been blessed and cursed. My mother left when I was a young baby, so this is my destiny,” Quick replied.

“Hold on, can I say a few words?” Janice interrupted.

“Sure,” Quick and Laura responded at the same time.

“Well, first and foremost, I also blame my father for introducing Akbar to that heist,” Janice said.

“Well, my dad had a chance to say no?” Quick replied.

“Yes, he had the option to say no, even during the time of the driver's killing, “ Laura intervened.

“My father picked up where he left off, bringing attention to this family even though he went back to school. He sold those computer chips after being released,” Quick explained.

After Laura listened to Quick and Janice's vent, Laura pushed her seat back, made her way toward Quick, and held his hand.

“I love you, Quick,” Laura assured him.

All three returned to their rooms, knowing they had to be ready.

On their way out, Janice held Quick's hand, kissed it, and said, “I am so proud of you.”

They hugged each other and headed in different directions. Before their yacht would dock, Quick wanted to ask Jasper a few questions. Quick stood in front of Jasper's room and knocked gently.

“Come in,” Jasper replied.

Quick walked inside Jasper's cabin.

“What is going on, young man?” Jasper asked.

“Well, my first question is about those diamonds. Were they part of the heist?” Quick inquired.

“How did you know about those diamonds?” Jasper asked.

"The officer Alvarez told Gus before he had been killed," Quick revealed his source.

Jasper looked away as though he wanted to avoid the conversation. Quick walked away, not wanting to go any deeper, but he knew that Jasper was aware of more than what he claimed.

Finally, their yacht made its way back toward Florida. The captain sounded like a small alarm, an old Exxon hand.

"Report to your location," he said, and they all stood in formation.

Laura had not slept the entire night. Janice, Jasper, and her few guests had been playing pool and drinking. Quick had been in his room preparing for his meeting with Vic. Gus had also been standing by if needed.

Quick grabbed his duffle bag and headed toward his family. Janice, Vic, and Laura had been Quick's only family. Janice still did not know about Akbar and how he died in a zip-line accident. Quick fixed himself; before going out, he clapped his hands together; that is when you knew Quick had been focused. As he exited the room, he could hear laughter coming from the lounge, so Quick went inside.

"Did you get enough rest?" Laura asked Quick while holding a hot cup of coffee.

Steam could be seen coming out of the mug. But instead of responding to Laura, Quick's eyes were finding Janice.

Janice waved, looking Quick directly in the eyes. Something did not seem right. As Janice approached Quick, she asked, "Are you ok?"

"Yes, I am fine," Quick responded.

"Gus should be waiting for you in California," he continued.

"How long will you be in Virginia?" Janice asked.

"Not long, love," Quick said as he kissed Janice on her cheek.

A voice blared through the speakers, "We will be docking in about 15 minutes. Please make sure to check for quarters before exiting the vessel. Hopefully, we will see you again soon."

Once the announcement was made, guests made their way toward the deck.

Quick gave a smile to Janice and said, "I love you, Janice." With this, they both went their ways as deckhands helped guests.

"Please be careful. Watch your step," one staff member gave instructions.

***

Quick made his way to the awaiting silver Mercedes GLC coupe SUV. Holding a sign, the driver stood there. The sign read, "Mr. Roland."

Quick approached the awaiting vehicle, and as he did, the driver said, "Hello, sir, I am Patty. I will be the driver," as she helped with his bags.

Quick was amazed at how beautiful she was. She had gorgeous hazel eyes and very attractive features. The young woman reached to open Quick's door, even though he tried to stop her.

"Do not worry about it," Quick said.

"Thank you. No one has ever done that," Patty smiled. She stood five feet tall, with long black hair dark complexion.

"I know your clients do not believe in chivalry," he giggled.

"They really don't. Most are stubborn rich men," she replied.

"I am not that kind," Quick assured as he sat in the car.

He immediately smelled a soothing scent of Chanel noir throughout the Mercedes-Benz. Suddenly a vibration of his phone went off; it scared him for a second.

"Hello?" He asked as he answered the phone call.

"Hey, young man, I will meet you in Virginia," Vic was on the other side.

Quick hung up the call.

"Sir, would you like me to stop anywhere?" Patty asked.

Quick looked at her in the rear-view mirror and said, "No."

Suddenly she turned around, with the words coming out of her mouth, "operation sheepskin."

Quick looked up, and a 40 caliber Glock handgun was pointed directly at his face. He tried to speak, but no words could come out of his mouth.

"Keep quiet," she said with one finger over her mouth.

She pulled the trigger; a single shot directly between Quick eyes. The young woman turned around in her seat and called an unknown number.

"Job complete," she said, hanging up.

She climbed into the backseat with Quick and his dead body going through his pockets. Afterward, she grabbed Quick's hand, cut off his thumb, and put it in a small box; her trophy kill.

Chapter 5: A New Seed Has Arrived

Five years had passed since Quick's body had been found in a vacant, abandoned small home in Florida with a single bullet wound to his forehead and his thumb missing.

Janice had finished college and was now doing her premedical at UCLA medical. Before he had been found dead, she had promised Quick that she would finish college.

“Mom,” a faint voice called out to Janice.

“What, Corey?”

Janice had been raising Corey with her mother. Corey was Quick's twin; he resembled his father so much. It bothered Janice that she never had the opportunity to share the joy of conceiving Corey on the same day as her birthday, the same day Quick had asked for her hand in marriage.

Janice called for Corey and gave him directions to grab his jacket.

“It is cold outside, Corey,” Olivia called out for him.

Olivia had been a longtime friend of Janice. She grabbed Corey's jacket for him and made their way downstairs. Janice had been waiting for Corey and her.

“Hello, Love!” Corey yelled as he walked in suddenly, running full speed and jumping into Janice's arms.

“Do not do that,” Janice explained, kissing Corey on his cheek, “I am not that strong,” she clarified.

Olivia busted out with laughter as the three made their way out. Once they were outside, Corey ran toward the Mercedes-C class coupe.

“Slow down,” Olivia screamed at Corey.

He looked back with a smile. Janice pushed her keypad, which unlocked the car and automatically started the engine. Olivia put Corey in the seat and buckled him in. Once inside, Janice looked over at Olivia and thanked her, “You're welcome,” Olivia replied, smiling as Janice backed out of her driveway, making her way down Ocean Blvd.

“Olivia, I am so happy that J.C. is coming home today. Especially out of the care of Dr. Dean; I still cannot believe he admitted J.C. into that hellhole. He even boosts a treadmill on electric shock therapy. It is just a blessing that J.C. has brain cells left after suffering through such a horrible experience,” Janice explained.

After driving for about 20 minutes, Janice made her way up to her mother's block in Naples, California. Her mother lived in a well-manicured area. Her home sat at the end of the block, surrounded by an 8-foot fence; she hated visitors just walking up to her front door without prior permission to the point that she installed surveillance cameras at the end of the corner of her block to see all incoming traffic.

Janice spoke to her virtual assistant, “Call mom.”

“Hello honey, I see you are coming up; just give me a minute,” she hung up the call, grabbed her Nike Air Max, and went out to meet her daughter and grandson, whom she loved dearly.

As Janice made her way to her mother's house, she could not keep her mind off J.C. She could not believe he was finally coming home from that horrible place. He needed them, not some hospital and electric therapy.

Pulling up the block toward her mother's home, she turned down the music and looked back; she noticed Corey had fallen asleep.

“Corey, Corey…,” Olivia whispered, “We are here at grandma's.”

Corey quickly woke up. Janice was there to drop him off at her mother’s.

After a hug and a kiss, Janice returned to her driver's seat, waving to her mother while pulling away. Corey stood waving and blowing kisses at Janice.

***

There was a knock on the door, “Now, who is it?”

“It is me, Virgil.”

“Come in,” Ms. Grayson replied.

“Joshua is asking about his cousin.”

“Okay, please let him know she will be here. Make sure you shower him up. He will be leaving today,” Ms. Grayson replied.

“Sure, will do. It is about time you release him back to his family.”

“Yes, the young man has been in dire need of his family.” Ms. Grayson mumbled, “Well, he should not have been here, to begin with.”

Virgil replied angrily, “What did you say?”

“Nothing,” Ms. Grayson replied.

Virgil walked toward J.C.'s room; Ms. Grayson followed behind closely. Exiting her office, she never noticed Gus seated outside her office door.

“Ms. Grayson!” Gus spoke out.

Ms. Grayson did not respond and kept walking at least three feet away, she turned around, and once she noticed Gus, she smiled with romance in her eyes, “Hello, Alan,” she greeted Gus.

Ms. Grayson had known Gus, going by the name of Alan.

“Well, hello, sweetie. I wanted to make it early for J.C.'s discharge,” Gus replied.

He grabbed Ms. Grayson into a passionate kiss, "We can handle this later. They are waiting for us," Ms. Grayson explained.

Walking toward the hospital conference room, its extremely waxed and buffed hallway floors looked like you were looking in a mirror.

Gus approached the hospital's elevator and asked, "So, your evaluation on Joshua is complete? Any problems with him being discharged?"

"No, no, not at all," Ms. Grayson replied, and at the same instance, she kissed Gus one more time before those hospital elevator doors opened.

Everyone knew that Joshua should have never been admitted into the psychiatric hospital. However, for some reason, Dr. Dean submitted a medical report to the children's court having him admitted.

Ding! Ding!

Ms. Grayson and Gus exited on the second floor, and there stood Dr. Dean, smoking in a non-smoking hospital. He was a real jerk, but no staff member would challenge him.

Ms. Grayson walked directly past Dr. Dean, entering a small hospital conference room with a huge leather conference chair.

Dr. Dean greeted them as the three positioned themselves, "Hello, Alan!"

"Hello, Mr. Dean," Gus replied.

"Give me a few minutes. Let me call Joshua," Dr. Dean said.

Requesting for Joshua to be brought to the camp room, after a few tipping rings, Virgil answered, "Hello, Dr. Dean."

"Virgil, can you please get Joshua ready?"

"Sure, Dr. Dean, he is ready. Give me a few minutes," Virgil hung up.

J.C. had been standing at the window, looking at Naomi. Naomi had been someone J.C. had become attached to, and he ended up falling in love with her.

"Aye, Joshua, you okay over there?"

"Yes, I am good," J.C. responded.

"Well, stupid Dr. Dean is downstairs waiting for you," Virgil informed him.

"Okay."

Joshua turned from the window not before taking one more look at Naomi as she sat in the hospital's well-manicured courtyard. Josh grabbed his Nike leather bag.

"Virgil, whenever you want a job, come holler at me," he said.

Virgil smiled, "Sure, kid, I cannot stand this place anyway."

As the two walked toward the hospital conference room, Joshua could not believe how long he had been waiting to walk out. Walking patiently up the hallway in silence, still having vivid images of his mother being killed, Joshua just wanted time for himself.

***

Janice and Olivia pulled into a parking spot just a few yards from the hospital entrance. Olivia, looking over at Janice, feeling the pain Janice has kept inside for so many years, tried to console her.

"Do not worry, Janice. Today is the day when J.C. will finally come home. You will now be able to start a new life with him."

Janice looked over at Olivia, "Thank you. You have been such a great friend, Olivia."

Olivia smiled at Janice, "You're welcome, Janice. I will always be here for you."

Janice and Olivia had been best friends since grade school. Olivia left California due to her stepfather molesting her at an early age; she ran away with her high school sweetheart and got married young in Tacoma, WA.

After a few years, her high school sweetheart got addicted to drugs and started beating her. She contacted Janice, needing help; Janice extended a hand to her, and Olivia had been with her ever since.

As the two walked through the hospital glass doors, a young blond-haired girl sat at the receptionist's desk; Olivia walked toward her.

"Hello, Miss. We have an appointment with Dr. Dean at 10 A.M.," Olivia informed.

"Okay, one minute, please," looking at her computer, she replied, "Yes, I see it. You are confirmed in room 236 on the second floor." She continued to give directions, "Elevators are to your left. Push the second floor, and 236 will be directly to your left."

"Thank you," Olivia replied.

Olivia and Janice made their way toward the elevators. They made it at the perfect time. As the elevator door opened, Janice pressed the number two. The elevator doors open on the second floor.

Ding! Ding!

Olivia and Janice walked toward room 236. As they both entered the conference room, Dr. Dean, Ms. Grayson, and Gus, known as Allen and Mr. Anderson, were waiting for them. Janice smiled as she noticed J.C. sitting at the far end of the table. Virgil was right by his side.

"Well, hello, ladies," Dr. Dean greeted and directed Olivia and Janice toward an empty conference room chair. But before they could take their seats, Gus stood up and pulled a chair back for Janice and Olivia each.

"Thank you," Janice and Olivia replied.

Ms. Grayson rolled her eyes when Gus assisted them with their chairs. Ms. Grayson had fallen deeply in love with Gus, not knowing he was a humanoid robot.

“Okay, now that we are here, let us get to business. We are here regarding Joshua Kemp's evaluation,” Dr. Dean looked toward Joshua and continued, “Well, today will be the day for your release from this facility.”

“Well, about time!” Joshua answered.

Everyone looked toward Dr. Dean. Ms. Grayson interrupted, slamming her hand on the desk. “Gentlemen, gentlemen! “

“Look, Joshua, we understand it has been such a long time, but today is your day.” Mr. Anderson stated, looking directly into Mr. Dean's eyes. He continued to speak, “You kept this young man here for so long,” Mr. Anderson complained.

“And let's not forget you put him on medication, knowing it should not have been prescribed,” Ms. Grayson added.

Dr. Dean understood exactly what Ms. Grayson was implying. All 100% true.

“Excuse me!” Joshua interrupted.

Ms. Grayson stopped dead in her speech.

Joshua reached down, picked up his Nike bag, and pulled out a bag full of Trifluoperazine pills on the conference room table. He just laughed while dumping the bag of pills.

“Dr. Dean, I have been holding on and saving these pills for this special day. I have not taken even one of these pills,” Joshua proudly said.

Dr. Dean stared in pure shock, pushed his chair back, and immediately left the conference room without finishing the evaluation meeting.

“Where are you going?” Ms. Grayson screamed with anger.

Dr. Dean turned around and said, "I am out of here! Take this paperwork and the kid off," slamming the door behind him.

Ms. Grayson smiled, knowing that Joshua had gotten him back at him at his own game.

"Well, can we go?" Janice asked.

"Yes, we are done here," Ms. Grayson answered.

As they walked out, Ms. Grayson grabbed Joshua by his hand and slid in a small piece of paper while whispering in his ear, "Feel they need to bleed."

Janice had noticed what Ms. Grayson had been doing but did not pay much attention to it. Gus and Joshua followed behind Janice and Olivia; Olivia looked at Joshua to reach the hospital exit.

"You are looking good, J.C.," she complimented him.

"Thank you, Olivia," J.C. responded.

J.C. approached Gus, "Can we visit Quick's gravesite?"

"Sure thing, kid," Gus replied.

Quick had done so much for J.C., Janice, and their family. He made sure Gus kept security on them as much as possible. Gus and J.C. walked toward a black two-door coupe convertible Bentley. J.C. could not believe Gus had so much swag for a humanoid; he fits right into society as a human being.

Gus pressed a small button on his keychain, automatically unlocking the doors and starting the two-door coupe. J.C. looked at Gus, still finding it hard to believe how real Gus looked.

As the two exited the hospital grounds, Gus started his playlist on the route to Quick's burial ground, playing Little Boosie's *I am sorry.*

J.C. leaned back, wishing Quick could see him now. Gus pressed a small black button, and the convertible retracted, letting their faces embrace the sunshine. Janice and Olivia made their way through the traffic.

“Damn, I am so proud that J.C. is back home; even having Gus around has been a blessing and a tremendous help to this family,” Janice spoke to Olivia, smiling.

“I am here for life,” Olivia assured her.

Janice was headed toward Orange County, LA.

“My father and Akbar were like brothers. Thirty-five million Dollars' worth of military software and computer chips had been stolen,” Janice opened up to Olivia.

Olivia was surprised and asked, “But thirty-five million?”

“Yes, thirty-five million!” Janice added, “Now that Corey is in my life, I need to find out more, knowing Quick would want that. He made sure he took care of my family and me,” Janice spoke as she pulled off the freeway toward the UCLA Medical Center, listening to slow jams.

About three miles in, Janice pulled in front of the UCLA Medical Center and said, “I am running late,” looking at the watch and exiting her vehicle. She told Olivia, “I will talk later.”

“Sure,” Olivia replied.

“Olivia, please don't forget to grab steaks, lobsters, and shrimp for tonight.”

“Sure, I will do that once I leave,” Olivia responded.

She climbed out, overtaking her position in the driver's seat.

Janice walked inside the UCLA Medical center's entrance and had one more look at Olivia waving.

“Good morning, Ms. Kemp,” security greeted Janice as she entered. He stood at six feet nine inches, muscular, with a bald head.

"Good morning!" Janice responded with a smile.

"Excuse me, Ms. Janice," a soft voice called out.

Janice looked around; a gorgeous woman stood with long black hair and the most gorgeous eyes she had ever seen. Janice looked directly into her eyes.

"Yes, can I help you?" Janice replied.

"Yes, you can. I just need a few minutes with you."

The woman fixed her hair, "Well, first, my name is... I know you are running late, but it is very important."

Janice scanned the hospital lobby for an open seat, pointing toward two open seats as they walked toward the lobby, taking a seat across from one another.

"Well, how can I start? I will start like this, I am originally from New Jersey, I moved to California at the age of fifteen, and during that time, I met a young man by the name of Clévon, but he eventually changed his name to Akbar."

Janice's eyes spread open, shocked at what she was hearing!

The woman continued, "I got accepted to the University of Hawaii and majored in computer science. To make the long story short, I had a chemistry teacher, Mr. Wong, he showed me how science plays a significant role in today's society. Just look at your life, working inside this hospital, saving lives, and helping people. It's never about memorizing textbook questions, but how you figure out the solution to your problems being an innovator."

"I am running late. Can you please let me know what you need from me?" Janice tried to cut it short.

"Okay, first and foremost, I am a dear friend of Akbar. I helped him develop artificial intelligence androids. We never imagined it would work out, but as we worked together and used resources, it all came together."

Janice noticed a clean-shaven gentleman looking at a magazine, giving her the eye.

"What is going on? Janice asked.

"Nothing. I see you are extremely cautious," she continued her conversation. "When Akbar came across the computer chips, he also came across the blueprints on how to develop humanoids. I am doubtful that Lee found out the individuals responsible for killing Akbar."

"Hold on, hold on! What do you mean killed Akbar?" Janice exclaimed.

"You did not know about Akbar?"

"No, I did not. This is my first time hearing about this."

"Well, if you did not hear about it, there must have been a reason. I am terribly sorry you had to find out like this." The woman replied.

So, this was the reason he never attended Quick's funeral, Janice thought to herself.

"After Quick and Akbar left for those few years, Akbar was ziplining, and his cord snapped. I hired an investigator to look deeper into the accident." Janice held her head high with a tear dropping from her eye as she listened to no name.

"I know, dear, it hurts me as well. More than you can imagine, John was my son!" Janice almost fainted when she found out Quick was her son.

Everyone called him Quick, but his name was John.

"I had to leave my son, my family had ties with the mob, and me getting pregnant by Akbar did not make it any better. So, for Quick and Akbar to be safe, I had to abandon him. Akbar still stayed in contact; John never knew. I followed him through his entire life." She confessed.

"Well, it will take me some time to process all of this. Can we talk later? Here is my number 310- 669-7788. Call me tomorrow; I am off!" explained Janice.

They both ended their conversation with a big hug. Walking away, Janice could not believe who she had just met; Quick's mother! She had been around all these years, a little anger inflamed inside her.

***

Gus pulled into Forest Lawn cemetery; J.C.'s facial expression changed to no emotion. He pulled his Bentley convertible behind a silver escalade; an older couple waved at them, returning to their vehicle. J.C. looked out into the distance.

"Are you okay?" Gus inquired.

"Yes, I am fine," J.C. assured him.

They both exited the car and headed to Quick's plot. J.C. looked down, taking a knee, holding his right hand on his forehead; tears came down his face. J.C. grabbed a track and field medal from his first pocket, something Quick had given him years ago.

Gus touched J.C. on the shoulder. "Trust me. Quick certainly would be proud of you guys. That is why Quick programmed me for you and Janice. Society has a misconception about artificial intelligence that we masquerade around to cause harm to humans. Scientists once said it would be unethical to build robots. But over the years, experimental roboticists who specialize in human robots said robots could be therapeutic," Gus elaborated.

"Gus, I do not know what would have happened to me without you. All those years coming to see me playing chess and teaching me about history surely has been therapeutic, and I want to learn more," J.C. said.

"Do not worry. That is why I am here for you. I will teach you as much as possible. I know Quick would not have wanted it any other way."

Gus and J.C. walked in the direction where the Bentley coupe had been parked.

"J.C., I need to tell you something I had not told you until this day," Gus said as they took their seats in his two-door Bentley Coupe.

"Okaayyy," J.C. replied as his heart rate started to increase.

"Your mother dated a man named Skip. He had her killed. Skip also deceived Akbar."

J.C. looks surprised as his mother had never mentioned this name all her life; she had been very private about her personal life.

"After killing a young woman on his boat, Skip disappeared. I have been keeping my ears to the streets, and it paid off," Gus mentioned.

A journalist named Todd McCord did a spread in the Los Angeles Times. The top ten restaurants to visit in Los Angeles. Gus reached into his pocket and pulled out a black book with a news clipping. There he was, standing tall with an alligator smile.

"That is Skip on the right and his younger sister. I guess after Quick and Akbar had been killed, he felt as though he could resurface," Gus revealed.

J.C. opened a pack of chewing gum, removed the wrapper, and put the gum in his mouth.

Looking into his eyes, Gus continued to explain, "With Laura's help, we have been able to find Skip's place," Gus revealed further, "With that devilish smile on his face, Skip sold everyone out for diamonds. Akbar never knew about those diamonds being a part of the heist."

"What a fu**ing a**hole," J.C. yelled out.

A young couple looked directly at J.C. "What is wrong? Never heard anyone swear and scream before?" J.C. yelled at the couple angrily.

Gus held J.C. by his shoulder, "Easy young man. You'll get plenty of opportunities to release all of that anger."

"Yeah, I am hands-on with that."

Gus looked at him and smiled, putting his Bentley coupe in drive and exiting the forest lawn.

"Quick had something incredibly special for you," Gus informed J.C.

Heading toward the 405 freeway, J.C. pressed the playlist, and "Dr. Dre" blared out through the speakers with the song *Places and Spaces I have been.* Gus hit the gas, and J.C. lay back, thinking of the most treacherous way of killing Skip and whomever else was responsible.

If it was not for that asshole, Skip, damn, mom! J.C. thought about it. Skip playing his mother, even getting her killed, after a short ride.

J.C. noticed the beautiful ocean-view homes. Gus made his way toward Laguna Beach. Being free felt like having the opportunity to get his hands on the people responsible for all this. The first person on his mind was Dr. Dean; secondly, Skip, even its order did not matter.

J.C. thought to himself, having flashbacks of his mother, who Janice loved so much. J.C. swapped out the daydream vision of his mother being scattered throughout their front lawn. Gus took an exit driving up the highway toward Laguna Beach. There was a beautiful two-story home.

J.C. looked up, "Wow, if it is not one thing, it's another," he said.

"Yes, this home belongs to you."

Pulling into the garage, J.C. looked around. He noticed a white 600 Mercedes.

"That is yours as well. Everything here belongs to you."

Gus parked as they both exited, walking toward a single black door.

Gus said, “Welcome home,” as the single door opened automatically.

They both walked into a two-person elevator closing behind them. Gus Pressed two on the elevator, taking them to the second-floor living room quarters. Upon reaching J.C.'s bedroom, located on the second floor, the elevator door opened, and “Have a lovely day” echoed throughout the elevator.

Gus and J.C. stepped into the massive bedroom from the elevator, and he noticed the ocean's blue water through a huge window. There was a King size bed and a 70-inch plasma smart T.V. J.C. walked around and noticed a massive walking closet that included a full body mirror, clothing, shoes, and accessories.

“J.C., let me show you around a little more,” Gus offered.

Exiting the bedroom, the two turned left down a hallway, “OK, J.C., there are three bedrooms up here, a shooting range, and a martial arts studio.”

As they continued walking, a stairwell led to the lower floors, a kitchen made of all stainless steel and appliances with island cabinets. They walked out, and Gus showed J.C. his living room.

“Take your time. Look around, and relax! There is plenty of food, don't eat much, though. Gus informed him that your cousin Janice is making dinner for you tonight,” Gus informed him.

J.C. walked away, still amazed at what Janice and Quick had provided for him, so thankful Gus had made sure Skip would be found.

Walking back toward Gus, “Hey, I am heading to my room,” he told Gus.

Once he walked in, paying attention to the room's interior, J.C. noticed a silver box sitting on the bed. He noticed his initials “J.C.” engraved in gold and picked up the box. There was a small button located on the left side. He pressed the small button, popping open the box.

In the contents was a single envelope addressed to Joshua Kemp. Two tears dropped from his eyes, picking up the letter.

It read, “Jay, seeing this letter is in your hands, this means that I have been killed in the process of finding them. Those responsible for killing your mother, my grandfather, and my dad. J.C. You are and always will be my little brother.

I also enclosed photos dated years ago; the people in these photos are responsible for killing our family members. You have to further my mission with the help of Gus. He has been programmed for you and the family.”

J.C. heard Gus walk in, and there he was, standing in the doorway.

“Don't worry, I have your back,” Gus replied.

J.C. reached into his pocket, pulling out a piece of paper. With Dr. Dean's address. Ms. Grayson had given it to him while leaving the hospital's conference room after being discharged.

“Gus, can you please take me here,” handing over Dr. Dean's address Gus reached for the paper looking at J.C.

“Sure, I will take you there.”

Closing the metal door, J.C. wanted to lie down, but Gus asked if he could show J.C. something before he could. Gus turned as J.C. followed him to the end of the hallway. A metal door J.C. had not noticed. He pressed his palm print. Then Gus instructed him to do the same, “sure,” J.C. complied. Suddenly, the door opened.

There was a cache of weapons, body armor electronics, AR15 assault rifles, etc. J.C. could not believe his eyes; the things he could not imagine in his lifetime.

“Everything in here belongs to you. I will teach you how to use everything in here,” J.C. looked around, surprised by all the tactical equipment. “Well, we have a few things to deal with, but first, get some rest. I will call you when I arrive.”

Chapter 6: Payback Is a Bitch

The sound of Miles Davis blared through the base speakers set in every room, and Dr. Dean continued cooking. The aroma of tomato sauce filled in Dr. Dean's home. Dr. Dean lived alone in his single-family home.

Ding! Ding! The doorbell rang.

"Damn it, who could that be?" Dr. Dean yelled out, talking to himself while tasting the tomato sauce.

Walking toward the living room, wiping his hands on his shirt.

"They are red. Bad to the bone." Dr. Dean said as Gus stood waiting patiently, opening the door without even asking who it was.

"Hello, may I help you?" Dr. Dean asked, handing him a small package; Gus waited for a response.

He looked at the printed name on the package.

"I do not know this person," suddenly, a punch came directly into Dr. Dean's throat, and he let out, going unconscious and hitting the floor.

Gus stepped over him, grabbed him by his wrist, and dragged him into the kitchen; Gus looked around and noticed he was in the middle of cooking. With his index finger, Gus tasted the red substance that had been simmering there and called J.C., letting him know that Dr. Dean was ready.

"I will be there in a few minutes," J.C. responded.

Dr. Dean began to moan; Gus immediately hit him again, knocking him out cold.

J.C. turned up the sound system in the black-on-black Jeep Cherokee, listening to Bryson Tiller's *Let em' Know.* He pushed the Jeep to one hundred MPH and smiled on the highway.

J.C. thought about all the horrible things Dr. Dean had done to him and every young woman admitted under his care. Naomi would feel different around her private areas after awakening when Dr. Dean drugged her food. Tears came down J.C.'s face when he thought about what Naomi had been through; today would be the end of Dr. Dean's horrible treatment of women.

Dr. Dean looked up with a blue vision; he could see Gus standing over him, pushing him to keep quiet. He just shut his eyes, not wanting to be hit again. All he could hear was Miles Davis playing throughout the house. All he could think about was *why? Who has sent this man to violate his home*?

After a few minutes, Gus slapped him across the face.

"Wake up, wake up, your time has come," Gus yelled at him.

After all those horrible things Dr. Dean had done, the time eventually turned around. The days, months, years, and years of abuse he inflicted on his patients will finally end today.

"I know you can hear me," Gus said.

Dr. Dean tried his best to keep his eyes shut. He could hear plastic. He peeped through his eyes and noticed Gus unrolling a body bag.

As Gus unwrapped and unzipped the body bag, Dr. Dean thought, *what the Fuck!* knowing today would be his last day on Earth.

As J.C. got close to Dr. Dean's home, exceeding the freeway, he could not imagine that this day had come. Dr. Dean's life would be in his hands in a few minutes. After all, the tragedy and being admitted to the facility had destroyed J.C.'s life. He could not wait for this; he pulled into Dr. Dean's garage.

J.C. felt nervous, but that feeling quickly passed away. He stepped out, walked toward his black Jeep's back, opened the hatch, and grabbed the duffel bags with all the equipment. Gus had instructed him on what to grab; a pair of surgical gloves, a jumpsuit, and a scalpel. He walked toward Dr. Dean's back door.

Gus had opened the back door; giving him a heads up, he pointed and told J.C. to put on the bodysuit and gloves before entering. Dr. Dean was lying in the middle of the kitchen floor, and Gus had zip-tied him.

Dr. Dean could not believe it was Joshua coming through the back door, entering his kitchen, and standing over him in a surgical setup, a surgical club.

"Please, young man, do not do this," Dr. Dean requested.

"Shut up!" J.C. shouted.

Dr. Dean began to beg and cry to let him go.

"You can cry all day, but it will not change your situation. All those years of hurting young women under your care, even admitting me under false pretense. Karma is a real bitch!" J.C. said while smiling.

Gus handed J.C. a needle with 20 milligrams of choline, typically used to relax muscles in surgery; Dr. Dean's eyes opened wide as if he had seen a ghost.

"Don't worry; this will be quick." J.C. injected him with the choline compound.

As Dr. Dean's muscles began to relax, his eyelids lay low. J.C. stepped on him, pulling down his pants and exposing his penis; with a scalpel in his left hand, ready to cut it off.

"This is for not knowing how to discipline your sexual desires." J.C. sliced Dr. Dean's penis, removing it from his body and placing it in the body bag next to him.

Gus grabbed Dean's arms, and J.C. grabbed both legs taking him to his bedroom inside the body bag. Dr. Dean was so relaxed, not even knowing that J.C. had cut off his penis. Gus instructed J.C. to remove his surgical suit and place it inside the bag.

"Don't worry. I will clean up the rest. You did very well, young man." Gus said.

"I guess it was all that built up aggression," J.C. replied.

"I will see you at home. This should not take long." Gus assured J.C. while sending him off.

J.C. exited Dr. Dean's home and felt relieved, knowing Naomi would be pleased. Finding Skip was next on J.C.'s list; then continue finding everyone else responsible for the destruction of his family.

"Go, J.C.," Gus instructed as J.C. pulled out of Dr. Dean's garage; a call had come in from Janice.

"Hello, cousin," Janice greeted.

"Hi, Janice," J.C. responded.

"So, what time will you and Gus arrive? I will be off in an hour." Janice explained.

"Give me about three hours, Janice," J.C. replied.

"Well, that is fine." Both hung up; J.C. started his vehicle and exited Dr. Dean's garage.

Dr. Dean lay inside a body bag in his bedroom. Gus could hear him choking inside the body bag. He had sprayed lighter fluid on the body. Gus lit a match, throwing an open flame on the body bag and lighting it up. Standing there looking at the fire, Gus smiled and walked away.

He returned to the kitchen, disconnected a gas line, and the smell of gas filled the kitchen. Before that gas could reach Dr. Dean's Bedroom, Gus exited the house from the kitchen's back door.

A parked Harley-Davidson was waiting for him. After a few minutes, Gus started his Harley-Davidson leaving Dr. Dean's home; he had just reached the corner of the street when suddenly, he heard a loud boom. Dr. Dean's house was in flames, and a nearby neighbor's home windows shattered due to the massive explosion!

"It is done!" Gus relayed the message to J.C.

J.C.'s phone vibrated; he looked down at his phone, retrieving the message. J.C. just smiled and headed toward Huntington Beach.

Emergency sirens could be heard in the distance as neighbors stood outside, in a state of shock, to see Dr. Dean's home in flames. Debris was scattered throughout the front lawn; Ms. Lewis stood behind the police. They had arrived before the fire department; Ms. Lewis could not believe Dr. Dean's home had been destroyed. Dr. Dean mainly stayed to himself, not wanting neighbors to work with him.

"Hello, Miss, can I ask you a few questions?" A young officer asked.

"I do not talk to the police. That is all I have to say." Ms. Lewis replied and walked away, fixing her scarf.

The officer stormed off, not looking back even once. The news channel 57 WXPYN had been en route.

Anita Ventura yelled, "What is taking so long to reach there?"

Anita Ventura had been working for Channel 57 for six years. Anita's cameraman was relatively new; Ortiz had been working for six months but had the attitude as if he had been a veteran worker.

As the Channel 57 News van pulled along Dr. Dean's St, orange flames coming from a pile of debris could be seen clearly. Neighbors were standing around, and the sun was about to set. People were so in disbelief that they weren't even paying attention to their surroundings.

Honk! Honk!

Ortiz pressed his horn so powerfully that he ended up hitting a neighbor of Dr. Dean's with his van.

“Watch out, asshole!” Ortiz hollered.

Anita looked over at Ortiz, loving his aggressive attitude. As he pulled toward the curb, the sound of a soda can crushed under the tire as Ortiz ran over it with the van's rear.

“Okay, how do I look?” asked Anita fixing her hair.

“You look excellent as always,” Anita and Ortiz exited the van grabbing their equipment to do a news report. During the explosion on Orange Court Ave, Ms. Lewis paid close attention; meanwhile, Anita searched for a location to do her report.

“This will be an excellent spot,” Anita said.

“Okay,” Ortiz responded.

As he positioned his camera, “5, 4, 3, 2, 1,” Ortiz counted down while giving Anita the thumbs up, making her aware that he had started recording.

Anita started broadcasting the news, “Hello, I am Anita Ventura, with channel 57 news team reporting live from Shine, Ville, California. Late this afternoon, an explosion destroyed a single-family home belonging to Dr. Dean, who worked at the Mission Psychiatric hospital. Emergency crews do not know if anyone had been inside. There was nothing left in this house. It would be a miracle if anyone survived this explosion.”

Ortiz zoomed in on the debris, pointing the camera back at Anita.

Anita walked toward a nearby neighbor and stopped her saying, “Hello, excuse me, miss. I am with channel 57. Could we ask you a few questions?”

“Well, sure, but just to let you know, I did not see anything, just like I told the police.” Ms. Lewis replied.

Anita Ventura looked back into the camera, smiling, “Well, thank you for this report. If we find out anything further, we will let you know.”

She added, “Tune back in. Thank you! Reporting live from Shine Ville, California, Anita Ventura.”

“How did I do? Your report sucked.” Ortiz responded.

Ending their short interview, Anita and Ortiz observed fire crews putting out the flames in the distance. A few hours had passed since J.C. had left Dr. Dean's home, having a great dinner with Janice and her family. J.C. sat at home looking over the picture of Skip, smiling and wondering how he could be so happy after he betrayed his mother and be responsible for who knows what else. J.C. summoned Gus to the conference room.

“Gus, can we go down to the restaurant? The one that Skip had a hand in,” asked J.C.

“Sure!” Gus responded, “Why not?”

“I will have a close friend set that appointment for tomorrow,” Gus informed J.C.

“Close friend?” J.C. questioned curiously.

“Yes, a close friend.”

He could not believe Gus could have a friend, but Akbar wanted Gus to be as normal as possible.

“Okay, sure, make that happen. We need to get a close-up on Skip.” J.C. said.

“Yes, we do!” Gus replied, exiting the room.

As the morning sun arrived, J.C. and Gus prepared themselves for meeting up in La Cienega with a close friend of Gus who was getting information on Skip's next big event and getting them invited. Gus discovered Skip would be hosting a fundraising party in Calabasas, California. The two drove down toward La Cienega to prepare for Skip's Fundraiser with smooth Jazz playing a song by Kenny G.

“We will have about 30 minutes to change our clothes,” Gus said.

“This is such a great idea passing as catering service, and the best part is Skip does not even know who I am,” J.C. replied.

“Just have to make sure Skip does not get away,” said Gus.

After about an hour's drive, Gus and J.C. finally arrived. It had been a gorgeous morning with the sun shining, and they could hear the birds chirping in the distance. Gus was parking his Bentley coupe when he noticed Patricia walking toward him with two catering uniforms, extending her hand and offering him a firm handshake.

“Hello, you two. I have two uniforms, and here are the keys to your van. Vic should already be there setting up the venue. Also, your clearance passes for the entrance are there too. Good luck, Fellas!” said Patricia.

J.C. and Gus began their journey toward Calabasas, CA. Skip had come out of hiding and become someone else; he even retired from being a customs agent. Skip's family had already been extraordinarily wealthy and liked being a part of the evil side. J.C. sat very patiently, knowing that today would be well worth it; he would finally be able to get revenge for his mother's death.

“J.C. does have more information on the task, for today we subdue Skip and retrieve those diamonds from his possession which had been manufactured in Singapore. Today a team of Russians will be involved in purchasing those diamonds; Skip has planned to sell them as official diamonds. However, you and I will throw a wrench in the entire plan.” Gus interpreted.

“Wow. This guy has many tricks in the bag. I am going to be on his ass!” J.C. got excited.

A call came in, “Hello.”

Vic was on the other side of the call.

“Hey Vic, we are getting close, about two exits away,” J.C. said.

"All right, the venue will be set when you arrive. Let them know you are with B.J. catering; they will direct you toward the service entrance toward the back of the kitchen."

"Sure thing," J.C. answered.

Hanging up on the phone with Vic, "Everything is set up for us." J.C. said while looking over and explaining to Gus. Gus pulled up to the security gate.

There stood a broad-shouldered young man in his early 20s, "Hello, Gentlemen. May I see your invitation?"

Gus produced paperwork showing the invite that B.J. catering had approved to enter, "Okay, gentlemen, you are cleared to enter. Follow that guy in the golf cart. He will guide you to the main entrance dock." The man told him.

Gus smiled, rolled up his window, and gave J.C. a thumbs up.

"Well, here we go, kid. Let the games begin!" J.C. sat there delighted with a devilish smile.

J.C. and Gus followed a young man who only looked 17. Their van followed up a narrow road, with only enough space for vehicles going in one direction; You could hear music playing at a distance. A song by Kevin Gates blared through the speakers, and J.C. could not believe rap music had been playing at such a prestigious event. Pulling to the rear of this massive home, J.C. still could not believe that Skip could live so lavishly at his mother's expense and that she had been killed because she was involved with Skip.

"You can park in that spot!" The young man bellowed out to Gus, pointing at him and J.C.

Gus gave the man the thumbs up, backing the van into position. J.C. had texted Vic, letting him know they had been parked and would be unloading their van. Vic texted back a small blueprint to Gus and J.C., where Skip would hold the meetings with the Russians. As J.C. and Gus exited, preparing to unload, a young woman approached, asking for their clearance permits. J.C. handed the permits over, thanking her.

To which she replied, “You're welcome.”

Walking away with a smile, Gus opened the rear doors and their catering van, removing food and all necessary supplies. With everything packed up, they began to walk through the central kitchen, making their way toward the courtyard. It had been centered in the middle of the home. The only way you could ever know the courtyard existed would be by entering the home.

“Wow, this home is a small fortress!” J.C. joyfully said to Gus.

They noticed Vic standing with a glass of wine in his hand, conversing with a young, attractive dark-complected woman. Vic noticed them as he held his wine glass in the air, walking a few feet toward them, and Gus started laughing.

“What's wrong?” J.C. asked.

Gus replied, “Well, every time I see this Vic guy in public, he is in the company of beautiful young women.”

J.C. began to laugh tremendously out loud.

“Excuse me!” touching the young woman, Vic introduced her to Gus and J.C.

“Hello! Nice to meet you both, gentlemen.” The woman replied.

The woman was Skip's sister. Skip had no idea that Vic had been creeping in late at night with her.

“Well, I will let you gentlemen get fully set up,” said the woman as she walked away.

“Who is she?” J.C. asked, looking at the woman.

“Well, she's Skip's sister.” Vic laughed.

“Surprised once again,” Gus answered, looking over at J.C.

“Well, enough about her. It's the time to get to business,” Vic said.

Gus and J.C. prepared themselves to make contact with Skip. They looked over the blueprints inside J.C.'s phone that he gave to Gus.

They both walked toward their target location. Once inside, J.C. noticed four nicely dressed men walking toward a meeting room where Skip would be meeting them. Gus and J.C. made their way toward the meeting room, but a worker stepped out of a side door before they could continue.

“Excuse me, gentlemen, you cannot be back here.” said the worker.

Gus felt the young woman could be a threat and immediately snapped her neck. He then pulled her back inside the room. Before he could get her body in, Skip came around the corner and noticed what had happened.

Skip looked directly into J.C.'s eyes, turned around, and ran off with a small briefcase.

J.C. immediately contacted Vic, “Mission has been aborted.”

Gus dropped the young woman, exiting their vehicle. Gus started the van and put it in drive mode, pushing down the gas pedal downward, not even noticing the golf cart he hit with the van, heading down the narrow road. He had no intentions of stopping.

Boom!!

Gus drove his van directly through the gate.

In shock, security could not believe what had just happened. One security personnel could be seen running down the narrow pathway as though he had a chance to stop the van.

J.C. looked over at Gus, surprised at how he did not hesitate to kill that young woman. Without saying a word, J.C. sent a text to Vic saying, ”you need to shut down and exit the property as soon as possible." Vic had already exited the property!

There was a secret exit that Skip's sister had told him about while Skip had been out of town months before. "Already on it." Vic smiled, sending J.C. a response.

J.C. dismantled his phone, putting it inside a small black zip-lock leather bag.

Gus drove back toward the staging area, where they would destroy their vehicle. Something went wrong for Gus to kill an innocent person like that. *It would be up to Vic on the next step for Gus!* J.C. thought to himself. Security refused to let visitors leave the mansion, and Calabasas police told them to lock down the facility. The only ones left were the Russians and Skip.

Vic understood he would have to shut the mission down for a while until he could access what the police knew. Vic drove up the highway. In deep thought, he realized this would bring their attention to the cameras, which would expose J.C. and Gus being there at the scene. Still not knowing why Gus would kill someone who possessed a threat. Years and years had passed, and today had been such a failure after getting so close to Skip!

Vic dialed J.C.'s number; after a few rings, J.C. finally answered, "Aye, Vic, What's up?"

"After you and Gus destroy that vehicle. Come by the shop first, before going home. We need to find out what the police have." Vic instructed.

"Okay, sure thing!" J.C. hung up the phone.

"Gus, after we destroy this van, Vic wants us to report to the shop," told J.C.

Gus looked over at J.C. and nodded; something just did not fit. Gus was not even saying a single word. J.C. had spent a year with Gus, and for him to not speak, J.C. understood that something had gone wrong in Gus's mind. Vic's phone vibrated; it had been Skip's sister, *damn!* Vic thought to himself while answering the call.

"Hello, love," Vic answered with a smirk.

Knowing deep inside he could not stand her, Vic would dispose of her if he could.

"Hello, darling, I was looking for you, but now I see you are no longer on the property." She replied.

"No, I left immediately after the commotion. I am on parole. Even though I am a business owner, I cannot have contact with the police!" said Vic.

"Well, it is okay. Quite a few people disappeared, even two of your employees," she said.

"Are you sure they worked for me?" Vic asked curiously.

"Yes, very sure, they both left in your company vehicle." she confidently replied.

On hearing that, Vic asked, "Why, what happened?"

"Well, a young woman has been killed, so now we are waiting for the police to arrive," she answered.

"I figured something horrible had happened; all I could hear was someone screaming and saying *call the police.* Then, you stormed out right after that, looking as though you had seen a ghost. So, I just packed up and left everything!" He explained.

She calmed Vic, "Do not worry about it. I will make sure you were never here, okay?"

Vic smiled as he had never expected her to do that for him.

Vic kissed her over the phone, "Thank you, maybe I will see you soon?"

“Sure,” they both hung up.

Skip's sister made her way toward security to retrieve all video surveillance. She had fallen deeply in love with Vic, not knowing he couldn't care less about her. Talking to a few security members, they followed as they made their way toward a guard station where a new hire had been working that day.

“Hello, Miss!” said the worker.

“Well, hello, young man!” Skip's sister replied, “Need all surveillance from today and restore it with something else.”

“Sure,” he said.

“Do not worry!” the new guard replied, “I will not tell the police you came for the surveillance tape,”

He took a piece of paper and pen, wrote $100,000 on it, due before Wednesday, and handed it to Skip's sister.

She looked and smiled, “Sure, what I need is worth it,” she laughed.

Then she immediately stopped, “do not spend it all in one place,” and she turned her back, walked out, and laughed.

He never knew how dangerous she could be and heard police rounding guests up—escorting them into one part of the mansion. Skip's sister looked at her two security team members; they had been with her team for over a decade.

She said, “Please take this surveillance tape to my safe! I have collateral damage to deal with!”

Wow! She thought to herself, another young woman killed on Skip's property. And here she was again, cleaning up his mess! Fixing her hair, she walked toward the detectives, looking directly at her path as if they knew her personally.

"Hello, miss, I am Detective Walker, and this is my partner Detective Gwen." said the detective.

"Hello, gentlemen. I am Patricia, the property owner." she introduced herself.

"Yes," replied the detectives.

"We are aware of that. One of your security team members directed us in your direction. So first, I would like to know the victim's name." asked one of the detectives.

"Well, her name is Kathy Ivanova. She was here working for us on a work visa. It is a shame that somebody would come to my fundraiser and kill her." Patricia said.

"Yes, it is a shame. Her neck had been snapped and broken from the looks of it. But to be 100% sure, the medical examiner will investigate further. Also, we will need your surveillance videos and guest list. We were told a van smashed through your security gate as well." said the detectives.

"Sure!" She replied to the detectives. "My help is at your disposal."

The detectives asked, "Who is this man?" looking at the guest's list.

As the detectives pointed out, there was Skip's photo on a guest flyer as being the host.

"Well, that is my brother Skip. He had left, and I do not know why," said Patricia.

Detective Gwenn rubbed his chin, wondering why she would admit that Skip had left, knowing someone had been killed on her property; it might be to make Skip look like a potential suspect.

"Yes, and about those surveillance videos, do I need a search warrant? Or could we just be fair and hand them over?" asked the detective.

“Well, sure, follow Harris, and he shall hand them over, Detective,” replied Patricia.

Following the security guard, the detectives walked away, and Patricia knew to call Skip immediately since he had just disappeared.

As she dialed Skip's number after a few rings, Skip finally picked up, “Hello, sis!” He asked, “How is everything?”

“Well, you should know, leaving me here to answer these questions with the detectives and your Russian friends,” replied Patricia.

“Are they still there? Why do those mother fucker's stay around? I have their money and never had a chance to turn over the remainder of the diamonds. Fuck! Fuck!” Skip yelled out.

“I am always cleaning up your messes every time,” said Patricia madly and then hung up on Skip.

If the police or even the federal government found Skip had been responsible for all of this and the heist. Akbar and Jasper served for a long time; they never once included Skip being the mastermind of this whole heist. Skip felt that people in his circle or those who knew him continued to die. Now, somebody he hired was murdered on his estate.

Skip began to replay the events before seeing the young woman murdered. The young man in the hallway looked familiar; taking a few breaths, he thought, *could that be Kay's son*? *They must know about me being a reason why his mother was killed.*

Skip reached into his pants, pulled out a nine-millimeter Taurus, and threw it on the passenger side floor of his Bentley coupe. He turned up the music with the song by Nipsey Hussle blaring out, *Hustle and Motivate.*

Chapter 7: Just a Little Longer

"Aurora, I'll always love you and your son," Janice said as they sat having afternoon iced coffee.

Aurora was fixed on Corey; he reminded her of Quick.

Janice noticed Aurora staring at Corey. "Yes, he is the spitting image of your son," she said.

She still could not wrap around her head that Quick's mother was sitting in front of her holding a conversation.

"Enjoying her firstborn son," said Janice as they ended their brief meeting and embraced each other, exchanging a kiss on the cheeks.

Later that evening, Vic had made it a point to shut Gus down for a moment. He would circulate another humanoid in place of Gus. However, Vic still could not completely understand why Gus killed that innocent woman, as he wasn't meant to kill innocent human beings. Vic knew that the law enforcement agencies would be extremely eager to find out who was responsible for what went down at Skip's party.

Why was a young woman murdered at this important event that hosted some of the world's wealthiest guests and important government officials? Vic would have to change locations by putting all plans in motion just in case law enforcement followed Gus through video surveillance traffic cameras.

Vic lived a double life; he was a family man, a highly skilled computer tech, and one of the best Harley-Davidson restoration mechanics in the country.

Rubbing his hands together and planning to put Gus on standby for a minute, at least until he could find out what law enforcement knew about the murder in Calabasas.

Vic thought to himself, *Skip has to be stopped immediately*!

Everything was clear now on how Skip was responsible for betraying Akbar and Quick's death. What hurts Vic the most was that Mr. Gibson had been killed, and he would not rest until Skip was found and punished for his part in the chaos.

Vic was unable to understand how anyone knew about Quick being in Florida. Vic started with the process of elimination; first, it would be Laura. She was like a mother to Quick. Laura had nothing to do with Quick's death. Quick being picked up from the pier in Florida, suddenly, a shock ran through Vic's body. "Jasper" whispered out of his mouth. Vic remembered a conversation between him and Jasper years ago.

"Akbar thinks he's a God's gift," Jasper had once told Vic.

Akbar had been awarded one of the country's best Software Engineer awards, so every software company wanted Akbar. Vic walked toward his war room, where Sebastian stood, waiting for his chance to be activated.

"Hello, Sebastian," Vic said, waiting for Sebastian to respond.

"Hello, Vic," Sebastian opened his eyes and said in a deep voice staring directly into Vic's eyes.

Standing at attention, listening for Vic's command, dressed in a pair of black cargo pants, a black turtleneck compression shirt, and black leather Nike boots. Vic continued to think about Jasper.

Looking him up, *where is Jasper*? He thought.

"Damn!" Vic shouted out loud.

That's it, he thought to himself.

Jasper had a 50% stake in Akbar's Company, and with Quick out of the way, Jasper gained it all. Vic programmed Jasper's home address into Sebastian's memory. He was sending Sebastian on the surveillance mission, knowing that Jasper had played a crucial role in Akbar's death. Vic called Net Jet to make reservations for Sebastian to fly out to Texas, as he owned a ranch home in Plano, TX.

"Hello, my name is Jeff. Welcome to net jet. How can I help you?" greeted the customer service agent.

"Hi, my name is Victor Mason. I would like to make reservations to Plano, TX."

"Okay, Sir, please allow me to place you on a brief hold," as soft jazz played, Grover Washington through the phone.

While Vic held on to the phone, the young man returned, "Sorry for the wait…."

The young man inquired, "May I have your account number?"

Vic answered, "Sure, it is 619164 expiration date. 2027."

"Okay, thank you," the net jet customer service agent responded.

"Well, Mr. Mason, we have a flight at 10:30 A.M., 6 P.M., and 10 P.M."

"Alright, I'll take your 6 P.M. flight going out," said Vic.

"Okay, Mr. Mason, I'll have you booked for 6 P.M. Please arrive one hour earlier than the flying time." The net jet customer service agent said.

"Okay, will do," Vic replied.

Vic and Sebastian had a brief meeting, explaining to Sebastian to get as much information from Jasper as possible, knowing that Jasper was responsible for the murder of Akbar, Quick, and Mr. Gibson. Certainly, there was no second-guessing about Jasper; Vic was sure about his feelings.

Sebastian would find out everything he needed to know about who was responsible while Gus was out of service. At least until the heat cooled down, knowing that the police would begin questioning everyone at the party, especially those who worked for the catering service; the only one who saw J.C. and Gus in the face had been Skip.

***

Suddenly, Officer Nelson approached detective Wallace's office, “Knock, knock!”

“Come in, Officer Nelson.” Detective Wallace responded.

He walked into Detective Wallace's office, carrying a brown folder and closed information, cross reference pending investigation on the murder on Skip's yacht out in Marina del Rey of a young Russian woman, finding it unusual that two women had died on the property owned by Skip.

“Detective Wallace, you might want to look over this,” Officer Nelson said.

While grabbing the case file, Detective Wallace said, “Thank you, Officer Nelson!”

He turned and walked away.

Detective Wallace grabbed his hot cup of coffee and preceded to open the case file given to him. He leaned back in his luxury leather seat, something that could be seen in a corporate boardroom.

Detective Wallace opened the case file and read about Patricia Hobbs, 29. He placed his hands behind his head and thought to himself; *a Russian named Patricia Hobbs? Born in Seattle, WA?* He thought maybe she had been working for an escort service. He dialed 213-692-4890, the Los Angeles County homicide department, to one of his old friends, detective Lawrence Gibbons, who served in the Marines with him.

After a few rings, the call was answered, “This is Detective Lawrence. How may I help you?”

“Old friend here, how's it going? This is Detective Wallace.” He replied.

“Well, well, well, how are the rich and the famous?”

Since Detective Wallace had been working in Calabasas, Detective Lawrence Gibbons joked with him.

“Everything is good this way. I'm calling concerning a cold case in Marina del Rey a few years ago, where a young Russian woman was found stuffed in a body bag and stored in the yacht’s storage freezer. I'm doing a cross-reference because a few days ago, another woman was murdered. The yacht is owned by the same person who owns this property here in Calabasas; I just wanted to check in and see if we had any serial killers on our hands. The yacht's owner was Steve McGee; his friends call him Skip, and his alibi checked out airtight, so this case remained cold. This time he is nowhere to be found. We have not been able to contact him, and we have surveillance of him speeding away from the scene,” Detective Wallace explained.

A couple of guests noticed him leaving, looking extremely nervous, finding it strange during his event as he had hosted a very important fundraiser.

“Yeah,” Gibbons replied.

“I checked every transportation service, airlines, buses, trains; nothing. The guy just vanished.” Detective Wallace told.

“Well, I'll make my way down toward the Marina del Rey and conduct a follow-up investigation to see if there are any new leads or if this guy Skip has been seen in the area.” Detective Lawrence replied.

“Okay, sure. Thank you, detective,” Wallace responded.

“Hold on before you go. Can you send me some of your reports?” Lawrence asked.

"No problem, I'll email you my Marina del Rey killing report around 10:30 tomorrow morning," Wallace said.

"Sure, no problem" they both hung up.

Detective Gibbons grabbed his 16-ounce cup of coffee and sat back, thinking about the two women being murdered and how Skip could be an ego-driven serial killer slipping up and losing control after killing a young woman on his home's property.

Beep! Beep! He looked at his laptop; he had received an attachment.

Detective Wallace had forwarded Detective Lawrence Gibbons his report; *wow!* Detective Gibbons thought to himself. *He works fast*; he was not expecting to receive the reports until tomorrow morning.

On the other hand, Detective Wallace pressed and opened the file. There were names of witnesses, including Eleanor Wentworth, age 49, who claimed she did not hear anything. Witness number 2, Donald Rand, age 64, also said he didn't hear anything. Witness number 3, Steve Boldman, age 75, and his wife Beverly Boldman, all alleged to hear nothing.

Damn!

Detective Wallace thought to himself, *he must have somehow used these people as see and suspect.* Detective Wallace grabbed his jacket from behind his chair; he had been running late from picking up his foster children. Detective Wallace thought deeply, wondering why Skip would disappear. *These young women were murdered on his properties, one on the water and one on land.* But this time, it was different; he was nowhere to be found for questioning.

While exiting the Calabasas Police Department, walking toward his silver Dodge Magnum, he heard a voice say, "Excuse me."

A man was standing next to a Ford F-150, tall and slim, dressed in very casual dark blue jeans, a red polo shirt, and a pair of Saucony track shoes.

“Yes, how may I help you?” Detective Wallace asked the man.

“I figured you'd be looking for me.” Skip replied.

“You're correct, Skip. I just finished doing a background check on you. So, you've been a longshoreman for more than 20 years, and you've inherited your parents' estate, huh?”

“Yes,” Skip replied.

“Your sister owns one of the best restaurants in Southern California. Life has been good to you...but the only thing I want to know is why two women have been found dead on your property?” Detective Wallace asked.

Detective Wallace pointed toward his passenger door and said, “Skip, please open the door and sit inside,”

Wallace figured Skip wouldn't come inside so he could speak with him off the record on this mystery.

They both stared at one another in silence until the detective said, “Okay, first things first, what do we know about the assistant being murdered on your property?”

“Well, detective, I was in charge of the fundraiser but had an emergency business meeting with some business partners from China. So, yes, I vacated without notice. I was really busy; I heard the news of my assistant being murdered here. So, I'm here two days later; my best option was to come down and clear my name.”

“Well, I will need your business partners' names and numbers and your whereabouts,” Detective Wallace replied.

“Yes, Detective Wallace, all of my contacts are logged in. Hold on one moment.” Skip went through his contacts, pulling up his alibi for Detective Wallace.

"Secondly, it's puzzling that a woman had been killed years ago on your yacht, and the killer has never been found," Detective Wallace further explained.

"I see that yacht in Marina del Rey is registered in your name. Maybe we should go inside and put everything you know on record."

After about a minute, Skip looked over at Detective Wallace, "I can't go inside," Skip said.

He looked around, and suddenly without noticing, Skip punched Detective Wallace directly on the side of his face, knocking him out cold. He could hear the sound of a bone breaking in Detective Wallace's jaw and blood spattering throughout the vehicle.

Skip immediately ran toward a black-on-black Mercedes-Benz AMG, which had been waiting nearby all along. All you could hear were screeching tires driving away. A young rookie police officer noticed Skip running from Detective Wallace's Dodge Magnum, and he immediately noticed Detective Wallace slumped over on his steering wheel. Walking over, he saw blood all over the driver's side window.

"Detective Wallace! Detective Wallace!" the young officer yelled out.

There was no response from Detective Wallace as the officer opened the driver's side door pulling Wallace out of the car; he was unresponsive.

The young officer called into the dispatch office, "Officer down! Officer down! We need assistance in the parking lot of the Calabasas Police Department."

Numerous officers responded to Detective Wallace's vehicle.

The dispatcher responded, "Help is on the way. The paramedics are en route."

Detective Wallace pointed at the young officer's shirt pocket and his name badge; it read McLaren.

“Detective Wallace, can you hear me?” McLaren said.

Detective Wallace let out a moan as he opened his eyes to look into the blue sky. He made a hand gesture letting the young officer know he could hear him. He then reached up and grabbed McClaren's pen out of his shirt; he angrily started writing down a phone number and a short note to pick up his foster children.

He also wrote for McLaren to call Detective Lawrence at the Los Angeles Homicide Division.

“I am calling for Detective Wallace; sorry, he will not be able to make it down. Skip just assaulted him.”

Detective Wallace handed his note to McLaren. “Sure thing,” McLaren gave him a nod.

The paramedics arrived and began to take care of him; Detective Wallace's jaw was broken for sure. He laid back, staring into the sky, slowly closing his eyes, relaxing, and trying to control the throbbing pain that he was feeling. He punched his fist into the concrete out of anger.

“Easy, Champ!” McLaren said.

Looking up, a short female paramedic walked up, “Please step back, Sir.”

McLaren had been focused on Detective Wallace and didn't notice the paramedic arriving.

“Sir,” she said, “can you hear me? Hold up one finger for yes.”

Detective Wallace held up one finger, “Okay.”

She replied, “That's perfect. We do not want you falling asleep on us.”

She pulled out a small flashlight, pointing it into the detective's eyes and checking his pupils. A heavy-set blonde male paramedic pulled up with a Gurney, lowering it to the ground. The paramedics put Detective Wallace on the Gurney and transferred him into the ambulance.

McLaren walked up to him and said, "I will make sure those messages are relayed."

He walked away with the writing pad in his hand.

Tapping Detective Wallace on his shoulder, one paramedic comforted him, "You'll be alright."

The paramedics secured Detective Wallace into the ambulance and closed the door.

One officer shouted in the distance, "That's for screwing someone's wife."

"Well, you cannot blame him if the wife let him screw," the short, stocky female officer replied and walked away laughing.

Detective Wallace lay inside the ambulance, "We got you, Detective!"

The female paramedic sat in the back with the detective.

The driver turned on the sirens and headed toward the hospital.

The paramedics called the Calabasas Hospital over their radio, "We have a black male in his mid-30s with a facial injury, possibly broken jaw, blood pressure, 127/80, pulse 69 beats per minute, our eta is approximately 11 minutes."

***

Later that evening, Vic had still been briefing Sebastian as he waited for J.C. to arrive, knowing his intuitive feeling about Jasper's motive. Vic knew this had to be addressed, although J.C. may not be open to looking for Jasper because he was Janice's father.

As Vic zoomed in on the security cameras, J.C. approached on his custom-built white-purple pearl-colored Harley Davidson built by Vic. He wore a custom leather jacket, cargo pants, and leather Chuck Taylor's. Vic could not believe how J.C. looked being out of the hospital. Being free, he looked so vibrant and alive; he had been through so much as a kid, waking up to see his mom killed in a car bombing.

As J.C. pulled up next to his old school 1957 Chevy Bel Air convertible, an ultimate classic, Vic noticed his facial expression as he looked serious and focused.

Vic walked toward Sebastian, “Well, he's finally here, Sebastian. He will be working side by side with you.”

Even though J.C. had no clue about him going out of town with Sebastian, he just knew about Gus and the years he spent with him. Vic pushed an oval button, which opened the door for J.C.

J.C. howled while removing his helmet, “Man, it is hot as hell out today!”

“Well, how do you know it's hot as hell?” Vic laughed.

“What's up, Vic? You said it's something you wanted to tell me?” J.C. asked.

“Yes, indeed, follow me first; I have someone I want you to see,” replied Vic.

As soon as J.C. followed Vic to the back room, he noticed a tall, muscular person; it was Sebastian looking at him and observing him.

J.C. saw Sebastian walking toward him and extended his hand to shake hands, “Are there any more surprises?”

J.C. replied, “There is one more, and it's about Jasper. I need you and Sebastian to visit him.”

“Vic, I already know about Jasper, and I am with you; everything you feel is correct. I have been feeling that way for a long while now.” J.C. said.

“After Akbar had been killed, Quick inherited all his assets while Jasper made awful decisions amidst being a bad gambler and lost tremendously. He was a major player; he was the only one who knew about Quick being in Florida and being picked up at the pier. Shortly after that, he disappeared for a little while. He did not even come to see his grandson." J.C. further explained.

"How did you know about this information?" Vic asked.

"Well, Vic, Gus explained everything to me, so there is a lot I do understand," J.C. replied.

"So, I have made reservations for you and Sebastian to visit Jasper in Plano, TX. You are heading out tonight. There is a duffel bag packed already. So many innocents loved ones have lost their lives behind Jasper and Skip's two-faced and scandalous ways. J.C., you know I would not make such strong accusations if there was no intense feeling deep inside." Vic made it clear. "We have not talked to this dude in more than a year. Five years!"

Vic's phone interrupted him in the middle of talking to J.C. He looked down at his phone; it was an unknown number. Vic knew exactly who it was.

He answered the phone, and a voice on the other end said, "We have a development. Detective Wallace called asking questions about the young girl killed in Marina Del Rey. Also, a new lead, Skip, turned up and assaulted Detective Wallace." The young woman's voice shared.

"Okay, thanks for the information. I will check into it," Vic replied.

After hanging up, Vic grabbed his laptop and immediately hacked into Calabasas security camera system. Vic looked around at all the surrounding areas on camera to see if he could get a visual of Skip. Well, gentlemen, Skip has surfaced. He was at the Calabasas Police Department, and the word is he has assaulted a police officer.

"Well, let me see if I can locate the vehicle he showed up in." J.C. walked toward Vic, wanting to see what Vic had been viewing.

There seemed as though Vic could burn a hole in his laptop. Vengeance had taken over his facial expressions.

"Well, we will deal with Skip on our return," J.C. replied.

Even though they had Sebastian and Sparta to deal with him, J.C. wanted to deal with Skip personally.

"You do not want Sebastian to deal with it?"

"No," J.C. replied, "I let him get away in Calabasas, not this time!"

J.C. smiled and rubbed his chin while speaking to Vic.

"This asshole belongs to me! How did you find out about him coming back to town?"

A longtime friend, who worked inside the LAPD and was an Officer of the Calabasas Police Department, called. They relayed a message to a homicide detective, and in doing so, he sent the name and the details of Skip being the suspect. Whatever information comes through, the department will get it first-hand. I made sure I had direct access.

"We also have a local newspaper reporter inside the Los Angeles Times," Vic said.

"Damn, Vic!" J.C. was very calculated and organized.

J.C. looked at Sebastian, "Well, big fella, it is time to get this show on the road."

Walking toward their duffel bag lying on a nearby table, Vic shut his laptop down and grabbed his keys.

"Well, let us get to business. My family is waiting on me, and it is getting late," Vic explained.

As the three continued their last-minute touch-ups for Sebastian and J.C.'s Flight, Vic and Sebastian loaded Vic's 1963 Toyota Land Rover. Vic had an exquisite taste for cars. J.C. had been on the phone with Naomi, assuring her he would return. He had been in love with Naomi since they were both hospitalized and spent years together at the mental hospital.

After all the abuse at the hands of Dr. Dean, J.C. promised Naomi he would handle it, which he did. Naomi smiled when she read the newspaper in which Dr. Dean's home had been destroyed in an explosion. She automatically knew J.C. was behind it.

After his business dealings out of town, J.C. bonded with Naomi. He had told her that he would come back for her one day. Those months away from Naomi had been very painful for J.C., but knowing she would be home beside him soon got him through the agonizing phase.

Vic screamed out, "Let's go, lover boy!"

Sebastian walked around, opening J.C.'s passenger door. J.C. looked surprised as he thought he should be driving.

"Nope, no," Vic said, smiling.

Vic adjusted his driver's seat, putting the vehicle in drive mode, and headed toward the Los Angeles International Airport. Vic played his music, and Anita Baker blared through the Land Rover sound system, playing, *You're my Angel.*

Sebastian sang along, and J.C. looked astonished. J.C. looked out his passenger window, looking at the passing car swaying his head from side to side while listening to the old-school songs being played by Vic. A vivid picture of Naomi appeared in his head.

***

Janice could not believe J.C. had just hung up on her. Whatever it was, she would question him in person. Janice still could not get Aurora out of her mind talking about Akbar and her making the arrangement.

Aurora walked out of Quick's life; it was still difficult to accept her apology. Janice knew Corey would never have to worry about her walking away. Janice walked over to Corey, picked him up, and held him close.

Corey reminded her of Quick as she kissed him on his forehead, laying him back down. As Janice stood over Corey, he lay there sleeping so peacefully.

Turning the lights off, Janice lay next to Corey. Fixing her goose-down feather pillow and setting her security system up, she immediately fell asleep. Janice did not notice Olivia looking over and smiling.

***

Looking into the distance, Santiago transfixed his sight on a hawk gliding through the clear blue sky.

“Sir, there's something important we need to discuss.”

“What's the problem?” Santiago responded.

“Well, Sir, Skip has brought attention to himself once again. In the first situation, we cleaned up, but another woman has been murdered on his property, and we just got information that he has assaulted a Calabasas police detective.”

“What?” Santiago hollowed, “What the hell? I've given him multiple chances to fall back. When did this take place?”

“Well, a few days ago.”

“What the fuck? I'm just hearing about this now?”

“Look, call Kemp, and let's find out where this guy is hiding at the moment. We have to find out before law enforcement and the ones responsible for killing Major Conroy. Because if we do not, operation sheepskin will be exposed. For years this has been a secret. The fire has died down.”

Santiago knew if Skip wasn't found soon, there could be serious consequences to his military operations; Operation Sheepskin would replace the nuclear arms race in the future, consisting of building humanoid artificial intelligence. It bothered Santiago that Jasper had never mentioned any diamonds being part of the heist.

Some things do not work out the way you want, Santiago thought to himself, *Jasper introduced me to this guy and got me into this mess.*

Santiago grabbed a bottle of black barrel tequila, looking out at his garden as the gardeners trimmed the hedges. He still could not believe his family had been murdered. He knew he could not be part of any conspiracy against the humanoid A.I. program, which should have been dismantled years ago. He sat in silence and orchestrated a plan for the Skip to be found.

***

Two gentlemen stepped out of a black-on-black four-door sedan, one standing about 6’3” and the other standing 5’9.” Both men exited the vehicle as quickly as possible.

“Damn!” Detective Gibbons yelled, “This must be some punishment for dealing with that married woman.”

He laughed aloud because he could not make any sense of why the air conditioner did not work.

“I have put in numerous maintenance slips; it's been two weeks. What's troubling me is the maintenance. Mechanic Craig said It would be done before summer.”

Detective Brandenton just smiled, knowing that detective Gibbons always had complaints, and that's probably why Craig said what he said. The mechanic was not in a hurry to fix his air conditioner. The two detectives walked toward the Marina del Rey pier at the gate. Detective Gibbons was so caught up in complaining that he almost walked into an older woman walking her dog.

“Excuse me,” said Detective Gibbons.

Making eye contact, he introduced himself and Detective Bradenton, "We are the Los Angeles homicide division. We are doing a follow-up on a woman found dead aboard a luxury yacht a few years ago. We are just doing a follow-up."

"Yes, I remember that young woman. The owner came down about a month ago. I guess he had been renting it out to a young couple. I docked a couple of lanes down from the Yorkshire yacht." She replied.

As they walked toward the boat, she looked over her backside as she knew Bradenton was looking at her, switching from left to right. They approached it and saw the name 'Deuce baby' written in black and gold letters on the rear of the yacht.

"Love the name of your yacht," Detective Gibbons praised.

"Thank you, that's my husband's favorite name. He hit big in a poker game and won over $2 million, his lucky charm. Even though we do not believe in luck, either you work hard for it or you don't. Follow me, gentlemen," she replied.

Soon the woman pulled out her keycard, and the three walked toward the gate. She scanned the key and opened the gate.

"Welcome, Bernice!" A strange voice came through the small speaker.

The three followed her onto the yacht and walked down below to a massive lounge containing a leather custom sectional, a bar, and a 60-inch plasma T.V.

"Would you gentlemen like a drink?" She offered.

"Yes, I will have a vodka dry," Detective Gibbons responded, "and my partner will have the same."

"Hold on! Hold on!" Detective Bradenton replied, "I will have Captain Morgan's rum."

Detective Gibbons looked over at Bradenton as the lady poured their drinks in. The two grabbed their drinks and pulled their barstools away from the bar.

"No, no, gentlemen!" She pointed toward her leather sectional.

They took their seat, and the woman started to explain what she noticed on the day of the murder of the Russian woman. "There had only been two gentlemen aboard, dressed real dandy. They stood about six feet tall and held a foreign accent."

"So, is that all?" asked one of them.

"No, I also wrote down the license plate number."

She handed the number to Detective Bradenton on a card; it said 'Healthy living food' on the front. Detective Gibbons smiled and shook his head.

"Well, thank you, and have a wonderful day."

They exited the yacht and walked away.

Detective Gibbons looked over at Bradenton and said, "Do you have another one?"

Detective Bradenton dialed headquarters after a few seconds, "Hello, Homicide Detective Ramirez."

"Hey Ramirez, this is Bradenton. Can you run a plate for me?" replied Bradenton.

"Yes, sure," responded Ramirez.

Detective Bradenton said, "The plate number is 913 XYY."

"Okay, Detective Bradenton. Hold one second, please."

The two detectives waited for the plate.

Ramirez returned and said, “Hello, Detective. The names came back with registered owner Cleavon Jackson, deceased over five years. I have an address, 681189 Shafter Lane.”

“Alright, thanks, Ramirez.”

“No problem, he replied, and Detective Bradenton hung up.

“You know what? I remember that guy. He had been nominated as one of the world's best software engineers and was killed in a ziplining accident. That address is where his company is located. Maybe we can swing by and speak with one of the employees.

I also remember he did time for manslaughter and had a conviction overturned. His dad had been one of the community's first African American grocery store owners. It is strange because his dad was killed during a traffic stop, and his son was found dead in Florida with a gunshot to his head. The sad part is that the girlfriend gave birth nine months later.” Explained Detective Gibbons.

“So sad. Well, let's get going. At least we have some wind in our face because this car is extremely hot.” said Detective Bradenton.

They pulled away from the Marina del Rey parking lot. Detective Bradenton began to brainstorm, wondering If Akbar had known who killed the young Russian girl. Or could he have been responsible? The Black four-door sedan pulled away, heading toward the 405 freeway. They merged through traffic, heading toward Akbar’s computer software company.

Both men enjoyed the cool breeze blowing in their faces. They both understood this was just the beginning; after all these years, one witness had been their only lead to the cold case. California sunshine glared through their sedan; Detective Bradenton pressed his electric window button and lowered the window.

Honk! Honk! A Toyota Camry hunk honks the horn.

"Damn! Everyone is in a hurry." Detective Gibbons yelled out.

Shaking his head, Detective Brandenton asked, "Look who's talking. How many accidents have you had?"

In the LAX accident, he ended up spending the entire year in the hospital, breaking both legs and numerous other bones. Still, he continued to complain about other drivers and whatever else he could complain about.

Detective Bradenton just sat back and enjoyed the ride. He remembered his case; he was the lead detective until he had to take leave. He had gone through a messy divorce, even to the point where he began to drink and miss work. He remembered receiving crime scene photos and the young girl having the word sheepskin carved on her back. He suddenly snapped out of his short-lived thinking because Detective Gibbons slammed on the brakes, almost rear-ending a motorcycle.

He pulled into the 10-story building; the parking lot seemed like every parking space had been taken. Detective Gibbons pulled directly into the handicapped parking, as this was the only space available.

"What? Are you going to park here?" Detective Bradenton asked.

Detective Gibbons started laughing as they exited the vehicle. He commented about being underappreciated because he noticed all the nice cars parked out front; he always felt that police never made enough money. Detective Bradenton just shook his head, not wanting to get into an argument with him. They approached the sliding glass doors while they looked around.

Sitting near the receptionist's desk, the young woman smiled, knowing they were visiting because they were the only people wearing ties; everyone else wore jeans and cool-fit shirts, very casual clothing.

"Hello," Detective Bradenton introduced himself as a detective, and so did Detective Gibbons, "we are the Los Angeles homicide division. We're doing a follow-up investigation, and we have information that the owner of this company might have been the last to see someone alive."

Detective Gibbons pulled out a picture of the deceased woman and asked, "Have you ever seen her before?"

The woman looked over the picture and said, "No, sorry, I cannot help you. I am fairly new here. But I can direct you to the director of the company. He is one of the founders."

"That's great," Detective Gibbons said, "because this case is over five years old."

The young woman made a call and, after a few rings, said, "Hello, Mr. Henderson. There are two homicide detectives here that are asking if they can speak with you."

"Of course, send them up," replied Mr. Henderson.

She hung up the phone.

"Okay, gentleman, Mr. Henderson will see you now. He's located on the 5th floor." The woman guided them.

She handed them a security key and said, "That's the only way you can operate the company elevator."

Detective Bradenton grabbed the key card and walked toward the elevator directly behind the receptionist's desk.

While Detective Gibbons looked around and said, "This building is gorgeous!"

They approached, swiped the card, and entered the lift. Pushing the number five, waiting patiently for the *'Ding.'*

"Welcome to the fifth floor." A voice came out of the small speaker located directly above their heads.

They exited the elevator and were met by a man standing there six feet, three inches tall, with a dark complexion.

"Hello, gentlemen."

They reached out to shake their hand.

"I am Mr. Henderson. How may I help you?"

"Is there somewhere private we can talk?" Detective Gibbons asked.

"Of course, gentleman," replied Mr. Henderson, "Follow me!"

The three men walked not too far from the elevator and approached a board meeting room.

"Please, gentlemen, have a seat."

They pulled their conference chairs away from the table.

Detective Bradenton cleared his throat, "Well, Mr. Henderson, we're doing a follow-up investigation on the murder that occurred over five years ago. Today we finally got information that the owner, Cleavon Jackson, was last seen leaving that area."

"Yes, we know he is deceased, but we're here to get closure on what happened."

Detective Givens pulled out a picture of the deceased woman.

"Have you ever seen her with Cleavon Jackson?" He asked.

"Well, gentlemen, the answer is no. Akbar was a very well-respected man and had been one of the most family-oriented men I've known. And I know he had history, but I assure you he is not capable of murdering anyone." Mr. Henderson replied with a little anger on his face.

Gibbons calmed him by saying, "Well, Mr. Henderson, we're not here to accuse him of killing anyone… just trying to put a few things in order because it's been over five years, and today is the first lead we've ever had. So, no disrespect to your longtime friend Cleavon Jackson Aka Akbar."

"Yes, I understand, gentlemen. I am sorry for the loss of the young woman myself, and I truly hope you will find the person responsible. We barely got over the loss of Mr. Gibson, Akbar, and his son. I've been working hard on the projects that Akbar wasn't able to accomplish due to his terrible accident in Saint Vincent." responded Mr. Henderson.

The three men pulled away from the conference table, but Mr. Henderson did not walk. Once they were out of sight, Mr. Henderson immediately called Vic.

After a few rings, Vic finally answered, "What's the deal, Henny?"

"Well, just to let you know, two detectives just came by asking questions about the murder in Marina del Rey." Mr. Henderson informed Vic.

"Wow! What the Fuck! last we heard, there were no witnesses." Vic yelled out.

"Well, I guess Skip probably talked to the police, but there is no benefit in all that because he's nowhere to be found," Henderson responded.

"Well, thanks for calling in, Henny," said Vic.

Vic arrived at LAX. He looked over at J.C. and informed him, "Well, the police just left Akbar's office asking questions about a murder that happened five years ago.

"Seems like, though the hunt is on and we're the lions and not the prey, Let's get it done! Make Mr. Gibson, Akbar, and Quick proud that their contributions to the community will never be forgotten because all they ever did was help those in need." Vic said.

"I truly understand, Vic. We will not stop until those who brought harm to our family are destroyed. And you know, I feel it because they murdered my mother. A single mom who worked two jobs to make sure I was okay! I spent years in a fucking mental hospital because of their decisions." J.C. added to the conversation.

Vic, J.C., and Sebastian sat in silence, focused on the task at hand, knowing that Jasper would put them a step closer to those responsible.

Even though Jasper was one of the main accused, what hurt the most was the betrayal by those close to the family.

So, after all, J.C. knew what Gus taught him; they should be loyal to the ones who genuinely care. During all the long talks with Naomi and Gus, J.C. wanted to be the best he could be for all those still living.

"Well, let us get going, fellows. We have a flight to catch." Sebastian grabbed his duffel bags with their gear as they made their way toward the Bombardier Challenger 650.

"Hello," a young woman greeted them, "Welcome to net jet. I will be your flight attendant. Please feel free to call me for anything you need."

"Hello, this is your captain. We will be flying into San Antonio, TX. Our arrival time is 9 P.M. Can you please fasten your seat belts until we reach cruising altitude. Once we do, like the wheel, you can release your seat belt. If you need anything, please feel free to call us, and we will have a flight attendant at your full disposal." The flight captain greeted them.

After a few hours in the air, Vic, J.C., and Sebastian finally made it to San Antonio, TX. Upon arrival, J.C. made a brief call to Janice, letting her know he was sorry for not making their family dinner that night. No one answered, and the call went to voicemail. Sebastian grabbed their duffel bags. Vic had been looking over Jasper in Plano, TX, and his ranch home.

"Okay, let's go," Vic said, smiling.

He could not wait to see Jasper's face, knowing that the game was up. It felt strange to kill Jasper; the agenda, for sure, would be to make him suffer as Jasper was everything to Janice. But it had to be done, all the horrible things he played a part in.

Vic looked over at J.C., knowing he was happy but at the same time sad that he and Janice were the only family members left and killing her dad was going to take away a part of her, but she had Corey.

The car pulled away from Lincoln Continental. Vic put on some jazz music, and Kenny G's *sentimental* blared throughout the vehicle.

Chapter 8: End Game

The alarm clock awakened J.C. at 4 A.M. It was time for his morning jog; today was a big day as Naomi was finally coming home. J.C. had promised he would never leave her again. He was torn once again because he tortured Jasper, but he also felt good as he got revenge for the gruesome killing of his mother and Quick.

J.C. could remember the terrified look on Jasper's face when he was awakened out of his sleep, and Sebastien grabbed him around his throat and picked him up with one arm. J.C. couldn't believe how strong Sebastian was.

Janice cried for weeks and did not even respond to his phone calls. He had only talked to her once since his return from Plano, TX. The first killing J.C. did was for Naomi for the things that Doctor Dean had done to her while in the mental hospital, but killing Jasper was different because J.C. knew him and the pain it would cause Janice, but it had to be done; now the task was finally Skip, but he was good at staying low. He was never seen again after the assault on detective Wallace. He had disappeared and vanished once again.

One thing that happened was that Jasper confessed to being in contact with a guy named Santiago. He said he was a significant part of everything that happened, and the most shocking part was that the entire ordeal wasn't behind the computer chips being stolen but diamonds that were a part of the heist, something Akbar knew nothing about.

J.C. remembered reading the letter Quick left behind, where he explained, and he recalled that Akbar was telling him that the computer chips were the reason behind the killings. But he came to find out they were dealing with a double-edged sword.

Those computer chips and someone else wanted the diamonds. So, it seemed like a never-ending story. J.C. thought to himself.

After his one-mile jog, J.C. returned to his Laguna Beach home driveway. In the distance, he noticed Vic parked in front of his brand-new Harley-Davidson Bucherer Blue Edition. This Harley-Davidson was worth more than modern D.E. Supercars. The ticket ran 2.4 million.

"Wow, Vic!" J.C. said with excitement. "This is one of the best yet."

Vic responded, "Yes, kid, someone old like me could ask for this beautiful piece of machinery."

Vic smiled, and the two made their way inside, heading toward the kitchen.

"Well, today is your big day. Finally, being able to bring Naomi home." Vic said with a smile.

"Yes, I can't wait! I'm not getting dressed. I'm going just like this!" J.C. smiled, looking at his watch.

"It's still early…so how about a quick game of chess?" J.C. asked.

"Sure, if you win, take the Harley-Davidson." Vic challenged.

"What, really?" J.C. asked with excitement, like a kid in the candy store.

Both men walked toward the chessboard located on a nearby table.

After about 30 minutes, J.C. yelled out, "Checkmate!"

Vic reached into his pocket, throwing the keys toward J.C.

"Well, time to go."

J.C. opened a nearby closet, grabbing two helmets. Not saying another word, J.C. walked out, jumped on the classic Harley-Davidson, put it in gear, and exited the Laguna Beach home.

Knock! Knock! Knock! "It is time. Your paperwork is done for your release." The young nurse explained.

Naomi could not wait after years of being confined in the dungeon of the mental hospital, even though she could go out to their courtyard and get fresh air, just the vibe of that horrible place. The worst part was being abused by Dr. Dean for years. They all knew! The only one who protected her had been J.C., even standing by his word on paying Doctor Dean a personal visit.

It brought joy to her spirits when she thought about it because no one had ever stood up for Naomi. So, J.C. would always be her special person because he was the only one she could depend on.

Packing her things and looking in the mirror, Naomi noticed a change in her face; she noticed happiness. What could it be that J.C. loved at that time? She brushed her hair one last time, grabbing her bag. A nearby nurse escorted her toward the waiting lobby where J.C. would be waiting for her.

"Hey, you!" J.C. yelled out, running up to her and picking her off her feet with a long passionate kiss.

"Damn, girl, you have gained a few pounds since I left." J.C. said while laughing, "It's okay! I love your thickness."

Walking out in front, J.C. handed her a helmet, "Here you go, Naomi, one of the helmets is for you."

Naomi looked at him and asked, "Are you serious?"

They both laughed. Naomi strapped her helmet on, "Well, let's get the fuck out of here."

She jumped on the back of the Harley-Davidson and wrapped her arms around J.C., squeezing him all so tight.

***

Vic felt a vibration in his pocket; grabbing his phone, it was Janice.

Vic answered the phone, "Hello Janice, what's going on?"

"Well, I just got a message saying they have Corey."

"What! Who has Corey?" Vic asked curiously.

"I don't know. The person on the other end stated that we had something they wanted. Please, Vic, I need Corey home."

Janice cried out, knowing if anyone could help, it would be Vic.

"Don't worry! Once J.C. gets back, we will both come over." They both hung up.

Vic thought to himself, *Damn! Could it be Skip? Santiago? But whoever took Corey just made themselves and everything they cared about a target. Wow! Corey is the last one to carry on the Akbar legacy.*

Vic walked toward the door; he noticed a Dodge Ram pulling up in the driveway. The last person who owned a truck like that was Akbar. Shaking his head, Vic grabbed his 45 calibers Smith and Wesson and cocked it. The car came to a stop. The door opened, and a muscular man stepped out from behind the door.

What the Fuck! Vic thought.

There stood Akbar in the flesh.

Made in the USA
Monee, IL
26 July 2023

ed211c78-e326-4ba3-a5fb-0d9e22a26c1dR01